Ultimately Understanding Biochemistry

AF601480

Jonathan Wolf Mueller

Ultimately Understanding Biochemistry

Jonathan Wolf Mueller
Department of Metabolism and Systems Science,
School of Medical Sciences
College of Medicine and Health
University of Birmingham
Birmingham, UK

ISBN 978-3-662-71888-9 ISBN 978-3-662-71889-6 (eBook)
https://doi.org/10.1007/978-3-662-71889-6

© The Editor(s) (if applicable) and The Author(s), under exclusive license to Springer-Verlag GmbH, DE, part of Springer Nature 2026

This work is subject to copyright. All rights are solely and exclusively licensed by the Publisher, whether the whole or part of the material is concerned, specifically the rights of translation, reprinting, reuse of illustrations, recitation, broadcasting, reproduction on microfilms or in any other physical way, and transmission or information storage and retrieval, electronic adaptation, computer software, or by similar or dissimilar methodology now known or hereafter developed.
The use of general descriptive names, registered names, trademarks, service marks, etc. in this publication does not imply, even in the absence of a specific statement, that such names are exempt from the relevant protective laws and regulations and therefore free for general use.
The publisher, the authors and the editors are safe to assume that the advice and information in this book are believed to be true and accurate at the date of publication. Neither the publisher nor the authors or the editors give a warranty, expressed or implied, with respect to the material contained herein or for any errors or omissions that may have been made. The publisher remains neutral with regard to jurisdictional claims in published maps and institutional affiliations.

Editorial Contact: Sarah Koch

This Springer imprint is published by the registered company Springer-Verlag GmbH, DE, part of Springer Nature.
The registered company address is: Heidelberger Platz 3, 14197 Berlin, Germany

If disposing of this product, please recycle the paper.

Dedicated to my teachers.
Dedicated to my students.

Foreword

This is an exceptional book, and so is this foreword. Here, Jonathan Wolf Mueller has made the commendable attempt to explain the basics of biochemistry in understandable words of everyday language. Biochemistry is a young science, relative to other disciplines, as it started off as an independent subject only in the 1950s. Biochemistry has developed at a spectacular pace ever since, culminating in the delivery of mRNA vaccines among many other things. Jon not only delivers biochemistry in simple terms. At many instances, he provides pioneering figures and schematics; always pairing chemical structures with images of everyday objects. The reader enters the world of molecules, for an exciting tour to the heights of the associated rules and the depths of the elementary processes. It is these rules and processes that represent *the* driving forces of evolution, resulting in increasing organization, differentiation, and diversification of life. Throughout the book, the personal commitment of the author is striking—sometimes it even seems as if he is entering a personal dialogue with the basic building blocks of life: with sugars, fats, nucleic acids, and amino acids, but also with proteins, enzymes, and DNA. Refreshing are the many metaphors, proverbs, and analogies with which he presents the complexity of the subject matter as simple, vivid, and understandable as possible. And the whole mixture is seasoned with biographical anecdotes and dry British humor, for that's where the author lives and works. In separate text boxes, he illuminates the backgrounds of individual phenomena, and on whatever is at hand—be it a beer mat or a napkin—he performs the odd magical trick with numbers, or two. Seemingly effortlessly, Jon transforms boring teaching subjects from traditional biochemistry lessons into a firework display of good ideas and sparking thoughts. He takes the freshman and the pre-educated by the hand, to show them the way through the technical labyrinth. He offers hints and shortcuts, and sideways. He creates an appetite for a deeper study of this exciting subject—ideally this opens the door to the "classic" textbooks of biochemistry. In this regard, I'm left with strongly recommending to the interested reader this book, that certainly deserved a title like "Understanding Biochemistry Playfully." May Jon find a large community of followers.

Werner Müller-Esterl
Frankfurt, in the Summer of 2022

Perception of This Book So Far

...an exciting book! Your book seems to be wild, but in a good way!
Prof. Simon Hammann, Stuttgart, Germany

A beautiful and exceptional book, indeed. Your approach certainly is unique.
Prof. Peter Bayer, Essen, Germany

A fabulous book! A humorous, visual, and very personal account of biochemistry. Many helpful analogies make the matter really catchy, and all the side stories repel boredom very effectively. Several passages really made me smile. Still, the book delivers the basics of biochemistry in a serious way. Highly commendable.
Hanna Alpermann, biochemistry student, Koblenz, Germany

A refreshing and adorable textbook: enjoyable to read, quickly to understand, and the many comparisons make it easier to remember, creating lasting learning pleasure!
Prof. Achim Gruber, Berlin, Germany

This book emanates a certain iridescence, not even starting to be taking the many pictures into account.
Prof. Joachim Rassow, Bochum, Germany

This is total joy for me, as I could resonate so well with your style. At times, your writing reminds me of Terry Pratchett. [...] A biochemistry book that is fun to read, finally.
Dr. Martin Reich, Berlin, Germany

Your book is wonderful, [...] first very funny, but then also witty and interesting...
Prof. Stan Kopriva, Cologne, Germany

In addition, these book reviews have been published:

Rassow, Joachim. **2024**. Buchrezension zu: Endlich Biochemie verstehen. *Biospektrum* 30, 125 (2024). ► https://doi.org/10.1007/s12268-024-2114-0.

Fischer, Ernst Peter. **2024**. Der fröhliche Biochemiker. Buchrezension zu: Endlich Biochemie verstehen. *Biologie in unserer Zeit*. 54(2):189. ► https://doi.org/10.11576/biuz-7206.

Navigation Pane

Nature is marvelous and the human body is astounding. A conventional introductory lecture can easily kill this fascination by yawning boredom; caused by abstract concepts and formulas, technical language and incomprehensible abbreviations. In this book, I provide a rough overview of various areas of biochemistry, without going into too much detail, attempting a balancing act. With this book, I want to champion an inclusive, low-barrier, and accessible biochemistry teaching delivery. Ideally, I want to preserve that fascination for Nature, that **WOW factor of life**.

The **idea for this book** comes from a masterclass lecture "Interested in the Biochemistry of Life?", which I have delivered to English school students in Birmingham between 2014 and the pandemic. I always enjoy outreach events like these ones for A-level students, as they offer me an opportunity to present what I find cool and exciting about biochemistry; without too much feeling the burden of a specific curriculum or an upcoming final exam. At the same time, I truly engage with **translating scientific concepts** into easier to understand terms, pictures, or analogies. Maybe useful for your next teaching sessions, or when writing a non-scientific summary for the general public, for some grant application.

An extension of the **concept of translation** started after the German version of this book, "*Endlich Biochemie verstehen*," was published in August 2023. I happily engaged to develop an English version of the book with my publisher, only to realize then how difficult it is to translate my scientific, but witty, sarcastic, and maybe ironic German text into English. I learned that being bilingual doesn't make you a good translator, yet. Can it be that I speak and think about identical biochemical concepts differently in various languages, well, just German and English for now… What you are looking at now is a thoroughly re-worked and carefully expanded version of the book that I published in 2023. Maybe not a translation at all, instead, I hopefully enhanced clarity, decluttered jargon even further, updated references, and strengthened some of my arguments. In line with this, I re-worked several figures and added two or three new pictures—one of them finally showcases TIM, the enzyme triose-phosphate isomerase, in full glory.

Several decisions had to be made. So, I opted for American Spelling, as I regularly publish scientific papers in American **Spelling**, our current *lingua franca*, after all, a "technical language." Sulfur certainly is spelled with the letter F. Wow, and then there are all the little battles about the spelling of individual technical terms. Together with the publisher, we aim for consistency. So, if you see something awkwardly spelled, and it is spelled like this a lot, it probably was meant to be like this. If it is a one-off, please please let us know. I conclude this book by providing a chapter on reading and writing scientific publications. Hopefully, this will help you navigating through the jungle of scientific publications.

Smart chemical reactions drive biochemistry, empowered by incredibly sophisticated molecular machines. The mechanism and the three-dimensional structure of these machines are becoming increasingly known these days. In this book, we discuss a selected set of reactions, and a few features of the enzymes involved. I have drawn all the **chemical formulas**, using the ChemDoodle software package. Many **illustrations** garnish the main text. I always pair up chemical structures with some kind of photo or picture as a memory anchor.

And then there are our **rough calculations** on a napkin. Many students of life sciences or health care professions seem to shy away from numbers. However, the napkins illustrate relations or proportions that are not always that obvious. At times, one might not have a napkin at hand, metaphorically speaking, so some of us might use the backside of an essay draft or an envelope. Fun fact: Germans have a strong relation to beermats, beer coasters, or *bierdeckels*, for doing all sorts of business, and rough calculations are one of them. Regardless of what support media you use, the idea of rough calculations is inspired by Ron Milo, Rehovot, Israel. I here provide a 2021 calculation from his team, as an example—our **napkin number I**.

Napkin I: How Many Viral Particles of SARS-CoV-2 Are There in the World?

Making biology as realistic as possible through rough calculations—that is one purpose of the napkins. These napkins should also soften the emotional barriers some students might have against calculations. *Math anxiety* is a well-known term among some teaching staff. Great calculations are available every now and then, for example, by Ron Milo and his co-workers.

Carefully compiled literature on the typical Covid-2 virus load in various body fluids and tissues indicates that an average person, at the peak of infection, carries about **10^9 to 10^{11} individual viral particles**.

Each Corona virion weighs about 1 femtogram; this is 1×10^{-15} g.

Hence, all virus particles per infected person are **1 µg to 100 µg**.

By early 2023, the World Health Organization (WHO) listed, for the whole pandemic, nearly **780 Mio**. registered cases.

All Corona viruses from all these people should weigh between **0.77 kg and 77 kg**.

We did not consider volume, but ALL viral particles from the whole Corona pandemic theoretically could fit into a wheely bin. I cannot imagine anyone who would not think of this scenario as a nice idea.

Here is more about these numbers—Sender et al. **2021**. The total number and mass of SARS-CoV-2 virions. *PNAS*. 118(25): e2024815118. ▶ https://doi.org/10.1073/pnas.2024815118.

For fans: The live Covid-19 WHO webpage—https://covid19.who.int/—provides figures on all registered Covid-19 cases. It was 771,407,825, on 22nd Oct 2023; and 779,034,940, on 9th Jan 2026. Kind of strange how little this number has increased. Is it that Covid really is a thing of the past? Or that we no longer test?

I reached out to Ron Milo to let him know that I'm now following in his footsteps, calculation wise. So good to be able to provide Ron's comment on these calculations, and on this entire book project as well—*This book looks great. I am impressed by Jon's style and knowledge. I wish this book much success and many more readers.*

Next are my science side stories, meant to be lighthearted and entertaining. These dedicated text boxes, **meanderings**, present additional information, useful anecdotes, or both. Some of this content is flagged up as **fun facts**—you may read it or not. Everything is in eternal flow. I want to convey that science never stands still. Yesterday's dogmas are constantly been tested in today's and tomorrow's research labs. Eventually, some of our hypotheses and beliefs are scrapped and replaced by new rules. Please note, just some of them. Establishing new concepts or ideas certainly isn't an easy endeavor. This is additional material that you don't need to read, but that might be fun to read. Please feel free to ignore entirely anything marked "fun fact". The **meandering text box** I put here as an example is about the flow of genetic information, and its comparison to music.

Close: ...and by the end of most chapters, there are the **C clauses**. They should encapsulate what was covered in the chapter. In addition, they might present a surprising and different angle to the topic. Certainly, they do not substitute for reading the whole text. It is **C clauses** for the many technical words starting with a C that the English language gives us. They are named for a silly German play of words that I won't repeat here.

Throughout this book, I will delve into how this molecule or that pathway could have come into existence in the first place. I propose **questions about the Why**, **the Where-from**, and **the When**, and I will even try to answer some of these questions. Please note that "Why questions" are somewhat problematic in biology. Do you know the quote about the drinking cup that is attributed to Werner Heisenberg? Within this book, we do not want to look too deeply into this cup. Instead, we will discuss insights from evolution, from prebiotic chemistry—that is the chemistry that might have prevailed on the early Earth—and from synthetic chemistry. These disciplines should give at least some sort of answers to the Big questions, loosely based on Richard Feynman's famous quote:

» "What I cannot create, I do not understand..." I quoted Richard Feynman from here—Way M. **2017**. What I cannot create, I do not understand. *J Cell Sci*; 130(18): 2941–2942. Editorial. ▸ https://doi.org/10.1242/jcs.209791.

ⓘ I really want to hear about **your opinions and thoughts**. Please get in touch via email, or social media. Which other scientific races would you like to be presented here? Do you know anything textual, witty, or prosaic that could feature in another edition of this book? Are there any noteworthy memes or memorable proverbs I have missed? All to make the next version of this book even better, in enabling and fostering playful and multimodal learning.

▪ An Actual Navigation Pane... with References to Other Chapters and Suggestions for Further Readings

Why the journal Nature Chemistry spells sulfur with the letter "F"—Editorial. **2009**. So long sulphur. *Nature Chem* 1, 333. ▸ https://doi.org/10.1038/nchem.301.

The more I do NOT actively look for them, the more I discover like-minded, beautifully written pieces of engaging biochemistry writing. Such as this one—

McCarty N. **2023**. Biology is a Burrito. Cells are so crowded, it's a wonder they work at all. Asimov Press. ► https://www.asimov.press/p/burrito-biology [accessed 5th April 2025].

Meandering: The Sound of the Genome

The **genome** can be thought of as the printed booklet with the score of a given piece of music. Nothing happens with this booklet alone. RNA and protein machinery (**transcriptome** and **proteome**) are like musicians and musical instruments; they interpret the whole piece in their own way. The resulting sound then is the **metabolome** or the phenotype.

An example within the analogy: The song "Satellite" was written by Julie Frost and John Gordon in 2007. People hardly took notice of this song. Only the musical interpretation by Stefan Raab and the spectacular performance by Lena Mayer-Landrut on May 29, 2010, in Oslo, Norway, made this song a winner song. Germany won the Eurovision Song Contest, only for the second time ever in the history of this competition.

Acknowledgments

This book started off as a mere translation of my 2023 German book "Ultimately understanding Biochemistry [German edition]," based on a publisher-provided raw-translation. The project, however, soon grew out of this. What you look at now is a completely reworked book, with many, many changes, hopefully to enhance clarity. I'm grateful for all the **feedback** I received for the 2023 version of this book. I incorporated as much as possible into the present book.

The book grew moderately in size, with a handful of new figures, or extensions to existing figures. You might want to regard it as some kind of a **new edition**—it just is that the publisher counts editions separately in different languages. Anyway, for this book, I checked and evaluated many, many sources. Some of them were the so-called "classic" **textbooks**; my most favorite one being the Voet & Voet, "my" biochemistry textbook, from back in the days. A prime example of how directly textbooks might have had an imprint on my way to learn, to teach, and to speak biochemically.

Quite a significant source of ideas and inspiration was what dear colleagues, old and new ones, have been sharing on **social networks**. Not too long ago, one could have said, they were "tweeting" about their successes, their experiences, or their memes. More recently, it is about a new connotation of the word blue-sky research, which previously plainly meant "basic research." I want to thank Simon Hammann and Leah Krevitt for allowing me to use their content, which I discovered at Blusky.

And then there were several **un-orthodox publishing items** that influenced this book: My contributions to the German *BioSpektrum* magazine, some for the British magazines *The Biochemist*, and *The Endocrinologist*, as well as the *ASBMB Today*, so called gray literature.

Oh, and then there is **Wikipedia**, the McDonald's of scientific communication. Every student and every member of staff uses Wikipedia, but somehow unreflected, at times. Regardless, without Wikipedia, I certainly would not have been able to refresh knowledge or to gain new access to different areas of knowledge in biochemistry in such a great detail and that quickly.

This book would not have come to completion without my editor Dr. Sarah Koch and my project manager Dr. Meike Barth, for their patience, their **professional support**, and their encouraging words, as well as their prolonged personal interest in the project.

Many **illustrations** are from my own photography archive. Others were provided by the publisher or from repositories. Some special illustrations, however, are from family and friends. Thank you to my nephew for letting me tinker with his Playmobil. Huge thanks to Angus Davison, Richard Wheeler, and the archive of the Max Planck Society for the kind provision of some exceptional illustrations.

Finally, I owe a big thank you to all my **proofreaders** Aryan Rolia, Brian Mintah, Christian Singh, Dominic Lai, Dumebi Ijeomah, Muskan Paul, and Sofia Denno, all current or former students, but also to colleagues, and friends from the

University of Birmingham, and beyond. Aryan Rolia, Brian Mintah, Christian Singh, Dominic Lai, Dumebi Ijeomah, Muskan Paul, and Sofia Denno I went through all my notes. This is the most recent, and hopefully accurate list of people who had been approached, and who also delivered meaningfully to this project. Dr. Christian Ludwig deserves special thanks for proofreading all the calculations on napkins.

Finally, big thanks again to Prof. Werner Müller-Esterl for his encouraging and supportive **Foreword**.

■ For Further Reading

Wikipedia shapes the perception of science to an unprecedented extent—Thompson & Hanley. **2018**. Science Is Shaped by Wikipedia: Evidence From a Randomized Control Trial. *MIT Sloan Research Paper* 5238–17, ► https://doi.org/10.2139/ssrn.3039505.

Disclaimer

I have researched and analyzed many references for this book, old ones and new ones. The text was carefully proofread by colleagues, some of my students, and friends. Nevertheless, I cannot guarantee the correct reproduction of particular facts.

This book builds on what my editor calls "my personal style." I sincerely hope you like my way of writing. It could easily be, however, that some marker or examiner would not approve this kind of biochemical narration, at all. Therefore, in an essay or an exam, please be mindful when directly quoting from this book with all its strange analogies and comparisons. I exclude liability for any incident or even exam failure caused therefrom.

Always remember the saying: *All analogies are wrong, but some of them are useful.* This slightly adapted quote is attributed to George Box.

Competing Interests The author has no competing interests to declare that are relevant to the content of this manuscript.

Contents

Making the Rules of the Game

Contents

© The Author(s), under exclusive license to Springer-Verlag GmbH, DE, part of Springer Nature 2026
J. W. Mueller, *Ultimately Understanding Biochemistry*, https://doi.org/10.1007/978-3-662-71889-6_1

1

There must be rules. And let's face it: Without rules, there is chaos. It is exciting to see how various biochemical processes "bend" or "interpret" these rules to their limits, apparently gaming the system. In tax law one might probably speak of "abuse of design", or of "stretching the law". It is almost a philosophical question, how Earth would be, if some of these rules were different, and nowadays, we can actually test such scenarios in computer simulations. Those "new" conditions could turn out to exist in the real world, maybe on another planet or within a highly specialized bacterium from a very exotic place—be it somewhere with lots of salt, most corrosive and alkaline soda, in the deepest depths of the World's oceans, or even in our guts.

1.1 Still Waters Run Deep

Introductory lectures in natural and life sciences tend to start with water (Fig. 1.1). As a student back in the day, I found that terribly boring. The older I get, however, the more I admire several truly unique properties of water. So, I will start here with water as well. Water is an astounding molecule, consisting of hydrogen and oxygen. Within the periodic table, oxygen is surrounded by carbon, nitrogen, fluorine, and sulfur. Comparing water to the hydrogen-containing compounds of all these other elements, CH_4, NH_3, HF and H_2S, reveals that water, and only H_2O, holds quite a few records:

- The melting and boiling points of water are very high.
- Water has an extremely high heat capacity.
- Obscenely high phase transition enthalpies are associated with melting and evaporation of ice, water, and water vapor, or steam.
- Water holds a record for the highest dielectric constant, amongst these substances.

Fig. 1.1 **Water is simply amazing**. Water is a record holder in quite some material properties. **Left**: Some pieces of frozen ice are depicted. **Center**: Nicely illuminated water droplets on a glass roof. **Right**: Steam over a pot of pasta. More about water and all its great properties in the main text. (Image credit: Ice, JWM 2021; Water droplets, JWM 2020; Steam, JWM 2006)

- Due to crazy amounts of hydrogen bonds, water has a very high surface tension.
- Finally, water can break down into two different ions and thus forms the basis for acid-base reactions and buffer systems.

But water isn't always that great. The density of water is high. This somewhat heavy fact, however, cannot be considered a strength, or even extraordinary as such. Instead, water's peculiar density is quite strange—it is also called water's **density anomaly**. Also, when it comes to **thermal conductivity**, water loses out to a lot of metals. But except these aspects, water is truly unique and forms the basis of all life as we know it. We will briefly expand on some of these points in the following.

Water displays an enormous **heat capacity**, meaning large amounts of energy are needed to heat water up. These energies are released again when water cools down. This property of water has led to heat itself being defined by water: One calorie—1 cal—is the amount of heat that can warm up 1 gram of water by 1 degree Celsius. Obscenely high amounts of energy are involved in the phase transitions of water—when ice melts or water freezes just as much as when water evaporates, or steam condenses. That's exactly why your gin-and-tonic stays so nicely cold for so long with just a few ice cubes in it (◘ Fig. 1.1).

A clunky chunk of aluminum expands when it is heated up, and contracts again when cooling down—just like other things in biology. Water, however, behaves differently. When water cools down, it first contracts—it becomes "heavier". At 4 °C, water reaches its highest density. But then it expands again at colder temperatures. Water doesn't do this by choice. At low temperatures, more and more regular hydrogen bridges form between water molecules, and these elaborate structures need space. Ice is about 10% lighter than cold water and therefore floats on it. This is very handy for the Arctic, where huge masses of ice float on the water—or used to float on it. Depending on when you read this text, we may have started calling the region around the North Pole of our planet the Arctic Ocean, instead of the Arctic. Water's **density anomaly** most literally is a "cool" characteristic. It prevents larger lakes from freezing down completely—in their depths, water at 4 °C will prevail all year round, so that fish can survive there in winter, they literally overwinter in deep waters.

Next, water has a huge **dielectric constant**. Water can shield opposite charges from each other very effectively. This makes water a super solvent for polar substances. Please think of the 300 grams of table salt that you can easily dissolve in one liter of water—if you really try, you might probably be able to dissolve 358 grams of sodium chloride in a liter of water, at room temperature, as an upper limit. Up to 300 grams of various biopolymers per liter are dissolved in water, all crammed together into living cells—what a strange coincidence. Without the special solvent water, this high concentration of various biomolecules might not be possible.

Water is special, even when broken down. Water can fall apart into two ions—two different parts with opposing charges. One is the hydroxyl ion OH^-, the other is a proton, a tiny, positively charged part of a water molecule, leaving the larger part of the water molecule and the negative charge behind. **But what is a proton, actually?** Physicists call this positive particle a proton; technically it is a nuclear

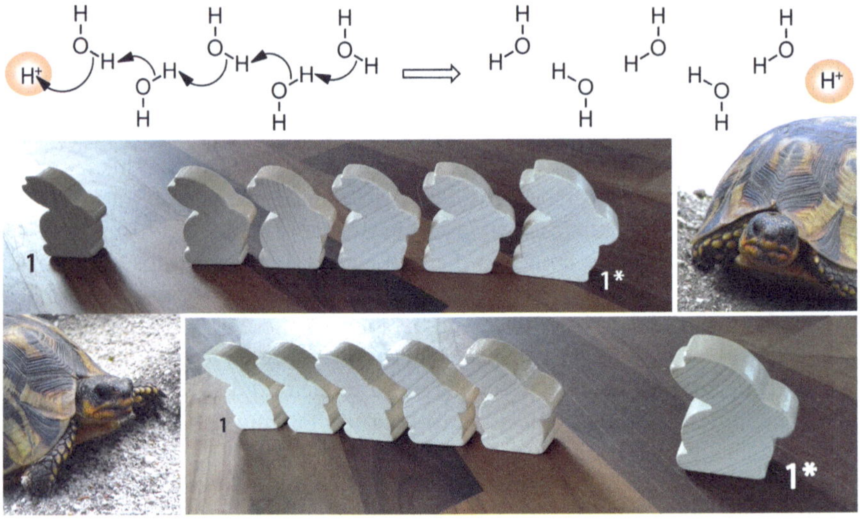

Fig. 1.2 **Protons bouncing around**. According to the mechanism of Grotthuss, protons "move" as if they are in the story of the Tortoise and the Hare. The figure shows jumping protons, comparing them to the hare. Some biochemical reactions, in which a proton plays a role, would not happen without bouncing protons; some other proton migration reactions become insanely fast. Samuel Cukierman beautifully covers the story of the German-born Lithuanian Theodor Grotthuss and his theory of proton migration—Cukierman. **2005**. Et tu, Grotthuss! and other unfinished stories. *Biochim Biophys Acta*. 1757(8): 876–85. ▶ https://doi.org/10.1016/j.bbabio.2005.12.001. (Image credits: ▶ stock.adobe.com; Bunnies and tortoise, JWM, 2022)

particle and no longer anything that is similar to an atom. The tiny thing shows properties from the quantum world, with wave-like properties. At times, a proton can tunnel through an energy mountain during a chemical reaction. Crazy, isn't it?

Such a proton does not like to be alone. Chemically, H^+ can perform quite a few tricks as well. As long as it is surrounded by water, it exists as the water adduct H_3O^+, the hydronium ion, or even in larger clusters of $H_9O_4^+$ or even more. Being bonded this way doesn't mean being restricted; within water, a proton can jump back and forth vividly (Fig. 1.2). This is because the severing and re-forming of bonds between water molecules are energetically identical. This bouncing of this particle rapidly accelerates the diffusion of protons in water. It is extremely hard to catch a proton; or even to isolate it… To the author, a proton is distantly similar to Schrödinger's cat. You don't know where it is… Nor how it is…

Capacity Water can absorb enormous amounts of heat and release them again. Because of water's density anomaly, deep lakes never completely freeze down to the bottom, so that life can hibernate. Water is a fantastic solvent for large amounts of polar salts and biomolecules. Ultimately, it can split into hydroxyl ions and protons. What great stuff water is. If water did not exist, it would be necessary to invent It. There is much more talk of protons, and of them running into buffers, in the next chapter…

Navigation

Icebergs are often depicted incorrectly. Nine out of ten parts of ice are below the water surface. Here is an online tool for practice. Draw an iceberg and see how it floats and when it turns ▶ https://joshdata.me/iceberger.html [viewed on 3.6.2022]. Then, draw motivational icebergs never again as upright floating icicles, please.

1.2 When Life Gives You Lemons...—The pH Scale and Buffer Systems

Whatever those sweet, little protons may be, we can count them in a watery solution very accurately. Differences in proton concentrations in the biological sample of choice, and in a reference electrode cause an electrical voltage difference, can be converted into the pH value, of course, after careful calibration of the pH meter.

The pH value represents the concentration of protons.

Please read elsewhere much more about how the pH value is calculated, here we only note that "pH" traditionally stands for the "potential of hydrogen", or "power of hydrogen", or Latin versions thereof. The pH value represents the acidity or alkalinity of an aqueous solution. If protons are here, but not there, a proton gradient is created. There is more about proton gradients in ▶ Sect. 5.7, when we discuss the F_1F_O-ATPase enzyme. Traditionally, the pH value is determined by using specific dyes. These are colored, organic substances that change their color

Meandering: The World's Best pH Indicator

Who hasn't dipped a stripe of a pH indicator into various liquids and watched the change of color. Does the name Litmus ring a bell here? Nowadays, pH strips have become more fancily looking—they feature several color patches, attached to a plastic strip, all to offer more accurate pH reporting, in different regions of the pH scale.

Back to the roots—well, or the leaves—of a much simpler pH indicator: the juice from red cabbage leaves. You could get an old and wilted outer leaf from a head of red cabbage, probably all year round from your local supermarket. Perhaps you could even get it for free, if you ask nicely. German folklore knows that red cabbage can change its color at times.

In northern German regions, this vegetable is known as *red cabbage*, which is a direct translation of "Rotkohl". In southern Germany, red cabbage is better known as "Blaukraut", which one can equate to *blue kraut*. Maybe you remember that Germans are called "Krauts" at times. A famous German tongue twister is about *blue kraut* and other stuff. Regardless, calling this vegetable red or blue cabbage is not just a question of geography, but also of acidity. In a non-representative survey, the author has looked at northern and southern German recipes for red cabbage. They

1

differ in color because whether or not generous amounts of vinegar or lemon juice are added.

The anthocyanins from within the cabbage are plant dyes that display splendid color changes, when the pH goes into the acidic or alkaline range. The figure in this **meandering** text box shows the structure of the anthocyanin cyanidin at a pH value of about 6 to 7. Red cabbage starts off as blue kraut then; it only turns into red cabbage when you add plenty of vinegar or lemon juice, because of the special properties of exactly those contained dyes, the anthocyanins.

Less known are the color changes of these anthocyanins in alkaline solutions, as the color of cyanidin changes over the entire pH scale. These *titrations* involve protons, moving back and forth. In alkaline solution, there even is the central ring with the oxygen opening up. All nicely suitable for chemistry experiments. For the kitchen, however, I do not recommend titrating red cabbage into a green ooze, or even a yellow mess; it isn't the tastiest thing. Trust me.

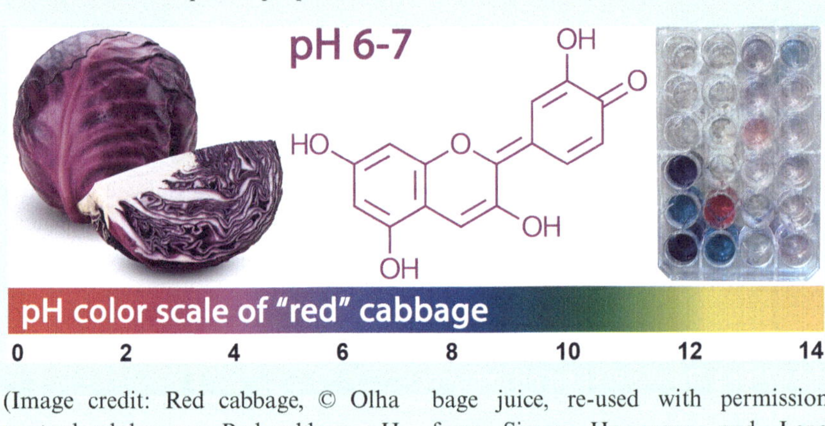

(Image credit: Red cabbage, © Olha ► stock.adobe.com; Red cabbage pH scale, JWM, 2022; Titration of red cabbage juice, re-used with permission from Simon Hammann and Lena Daumann)

upon at a change of the pH value. Personally, I think that simple red cabbage juice represents the best pH color indicator on the planet (see the **meandering text box** on a very special pH indicator).

With regard to any pH matters, a lot of biomolecules have their say, because they display both acidic and basic properties. Thus, these biomolecules can mitigate or absorb smaller changes in the pH value—we have arrived at biological buffers. I mention this here, because biological systems more often than not show a finely and elaborately regulated and stabilized pH value.

The pH value of human blood normally should be a constant 7.40, mainly maintained by the carbonic acid-bicarbonate buffering system. Should your blood pH ever fall below 7.35, this would already be referred to as acidosis, at times linked to ketosis by physicians or clinical chemists. Already that tiny shift in pH would make you not really feeling well anymore.

In the biochemical laboratory, we use buffer substances with rather strangely sounding names, such as Tris or MOPS. There is a **meandering text box** about the history of these buffers. Its title is "Good old times".

Meandering: "Good Old Times" or Why Modern Molecular Biology Exists at All

In the early days of biochemistry, phosphate-buffered saline (PBS) was in use, basically always and constantly. Today we rarely use PBS anymore, probably only when washing some mammalian cells that are cultured in a plastic dish. Instead, buffers in common use these days, come with strange names such as MES, HEPES, Tris or MOPS. These buffer substances, and a few more, are referred to as Good buffers. But this is not because they are "good" buffers. It is more around Norman Good, who chemically made and systematically characterized these compounds in the 1970s and 1980s. Good's buffers mainly contain organically bound amines and other protonatable groups—groups that can give or accept a proton. Most importantly, all these buffers hand over their protons in a pH range somewhere between pH 6 and 8. Phosphate buffer has its midpoint of buffer exchange, one of its pKA values, at the pH of 7.2. This means that Good buffers work approximately in the same pH range as phosphate buffers, but they are guaranteed free of phosphate.

Why is the absence of phosphate so important? For those of you who like a bit of drama: without the Good buffers, modern molecular biology would simply not exist. An absolutely annoying thing about phosphate solutions is that the divalent cations Mg^{2+} and Ca^{2+} are practically insoluble in them. Meaning, you can only have one thing at any place, at any given time—either a lot of phosphate OR higher amounts of magnesium or calcium. If you mix them, their barely water-soluble salts will precipitate immediately. Well, this circumstance is quite practical for our teeth and bones, which, to a great extent, are built of various calcium phosphates.

Back to the foundation years of molecular biology. A polymerase chain reaction (PCR) would never have succeeded in PBS. Kary Mullis is acknowledged to have invented the PCR. His first PCR reaction mixtures contained 10 mM free magnesium ions. Today, PCRs and many other molecular biology reactions run in buffers containing between 0.5 and 4 mM free Mg^{2+} ions, with the pH kept stable by Good buffers.

Of MOPS and Mops Fun fact: Pugs are a well-known breed of dogs, and a pug is called a "Mops" in German. So, the picture in this meandering shows the buffer substance MOPS (3-(N-Morpholino)-propane-sulfonic acid), one of the most commonly used Good buffers, paired with a "Mops", a pug. MOPS contains a ring structure with a nitrogen atom in it. This is where MOPS can buffer with a pK_A value of about 7.20—ideal for biochemical research. The pug in bronze is an absolute favorite of one of the Greatest German comedians, Loriot. His pugs-from-the-woods—his *Waldmöpse*—are truly classic. In honor of Loriot, bronze sculptures of *Waldmöpse* have been released all around the city of Brandenburg an der Havel, where he was born.

Back to the lab: Biochemists who know Loriot's humor might say—"Life without MOPS buffer is possible, but pointless."

(Photo: *Waldmops* in Brandenburg an der Havel, Thomas Müller, 2020, Reproduction with kind permission)

■ **For further reading**

A wonderful and comprehensive overview article about Good and his buffers is provided by Gary Pielak. It comes with the surprise that Tris is NOT a Good buffer, it was known before. And there are passionate Tris haters, among them Pielak himself.—Pielak. **2021**. Buffers, Especially the Good Kind. *Biochemistry*. 60(46): 3436–3440. Review. ► https://doi.org/10.1021/acs.biochem.1c00200.

Loriot is quoted according to his book "Dear Ladies and Gentlemen...", Diogenes Verlag, Zurich, **2005**

"It is difficult to make predictions, especially about the future" (attributed to Niels Bohr). However, if you mix water with some salts, a bit of some sort of acid, or some alkaline stuff, it would be cool to be able to predict the behavior of the resulting diluted buffer solution, at least somewhat quantitatively, to some extent. Because this was a thing in 1908, Lawrence Joseph Henderson wrote down an equation which allowed you to calculate the pH value of a buffer solution. Almost 10 years later, in 1917, Karl Albert Hasselbalch came along and re-shaped this equation into its logarithmic form—the **Henderson-Hasselbalch equation** was born. Why it is exactly this equation that made it into numerous textbooks and curricula remains unclear to this day.

$$pH = pK_A + \log_{10}\frac{\left[A^-\right]}{\left[HA\right]} \text{ or – that ratio turned around – also } pH = pK_A - \log_{10}\frac{\left[HA\right]}{\left[A^-\right]}$$

Admittedly, the Henderson-Hasselbalch equation looks somewhat daunting to many students at first glance, probably both because of its long name and its "mathematical" look. Nevertheless, this equation quite accurately describes what can be observed during an experimental titration at near-physiological pH values. You may ask what a titration actually is? Please check within the glossary of this book. The logarithmic form is necessary because the pH scale also is logarithmic. **Napkin II** highlights the two things that one can learn from the Henderson-Hasselbalch equation.

Napkin II: What Can the Henderson-Hasselbalch Equation Do for You?

Two biochemically related things can be learned here: (1) how buffers work quantitatively and (2) what the acid constant, also known as the pK_A value, says about a specific functional group. Let's start with the term:

"*log_{10} [acid] divided by [acid residue]*", where the bracketed terms mean "molar concentration of…"

In short, when an acid and its conjugate base (the acid residue) are present in a 100 to 1 ratio, then the pH value would be $pK_A - 2$

at 10 to 1, it is $pH = pK_A - 1$

at 1 to 1, the $pH = pK_A$

at 1 to 10, it is $pH = pK_A + 1$

at 1 to 100, it is $pH = pK_A + 2$

The logarithmic representation is to be commended here. Uncomfortable terms such as 100-to-1 or 1-to-100 are easily transformed into something cute such as "+2" or "–2". Otherwise, the pK_A value is the pH value, where there are equal amounts of acid and conjugate base in your solution.

The Henderson-Hasselbalch equation is fantastically suited to describe the behavior of buffers with pK_A values between 5 and 9. Above or below this range, this very equation more and more becomes a rough approximation, though.

With the Henderson-Hasselbalch equation, one can quantitatively understand how buffers work, at least in the physiologically relevant range: A buffer substance buffers most strongly at pH values that are close to its pK_A value; the buffer might still work one or two pH units above or below the pK_A; it loses buffering capacity beyond this range.

Put differently, at pH values in biological compartments, only those functional groups join the buffering game, whose pK_A values are close to the local pH value from within the compartment in focus. Any functional group with pK_A values two or even three pH units away can be assumed to be permanently charged or uncharged. In the everyday life of a biochemist, this means that it is mostly histidine and cysteine sidechains in proteins that are biologically interesting buffers. For the “charged” amino acids glutamate, aspartate, lysine and arginine it requires some biochemical “convincing” to become active buffers. Dedicated cellular compartments and active centers of enzymes traditionally are quite good in exactly this type of “convincing”. In such special surroundings, the pK_A values of selected amino acids can be dramatically different to their standard values.

Here is some background reading around the Henderson-Hasselbalch equation—Po & Senozan. **2001**. The Henderson-Hasselbalch equation: its history and limitations. *J Chemical Education*. 78(11) 1499. ▸ https://doi.org/10.1021/ed078p1499.

How to put the Henderson-Hasselbalch equation to best use? As an A-level student or high school graduate, I had to learn it by heart. Nowadays I present it to medical and life science students. I admire people who can derive this equation directly from the law of mass action. Looking at proteins or other large biomolecules, this equation provides the framework to you to better judge the importance of functional groups displaying a range of pK_A values. You may be able to guess which group “have a say” protonation-wise during catalysis, and which other groups remain lazily or passively in a certain protonation state.

After all, I cherish most the equation from Mr. Henderson and Mr. Hasselbalch for its usefulness in daily lab routine—at the pH meter. If you ever pH-ed a buffer yourself in the lab, you’ll know this: To adjust it as finely as the pH of our blood, is not easy. Trust me, you will make fewer mistakes while pH-ing, if you are aware of the pK_A value of your buffer.

Question What was your most remarkable pH titration so far? Write to me, if it really was exciting. Should your titration have been truly embarrassing, I would want to hear about it even more. I do want to hear from you. My own story is here, in a small **meandering text box** on pH titrations.

Meandering: Making a Fuss at the pH Meter

We all know this. You are at the pH meter. Lunch or after-work hours are seemingly close. You add more and more acid or base to your solution or reaction… **Nothing happens**. Feelings of frustration creep up. I, at least, would then start to increase the volume of acid or base per titration step… More is more, isn’t it? … but suddenly, the pH value of my buffer is far off—far above or far below. So, we enter the realm of back titration. Lunch or the end of the workday may disappear in the blurry distance.

Anyone who knows the acid or base value (affectionately called pK_A or pK_B value) for their buffer and also pragmatically understands the Henderson-Hasselbalch equation, might perform better at the pH meter, at least slightly.

Really. A good illustration is the making of an EDTA solution. EDTA is a chelator for metal ions; it is widely used in biochemistry. The free acid EDTA is uncharged; it is barely soluble in water. Trying to make a watery solution out of it will first result in a nasty milky slime. When titrated with base, the EDTA solution will quite suddenly become crystal-clear. How exciting.

Probably my most exciting pH titration was that of two small chaperone proteins—affectionately called Pin1 and Par14. In a very high-resolution crystal structure of one of these two proteins, we could see individual hydrogen atoms. They are displayed as red lattice in **this meandering's figure**. This was probably the only time in my life, I have directly seen hydrogen atoms experimentally. Using nuclear magnetic resonance spectroscopy (NMR), we then zoomed in at individual histidine side chains. Remarkably, the local pK_A values of these histidine residues changed upon the addition of a substrate. Even more exciting.

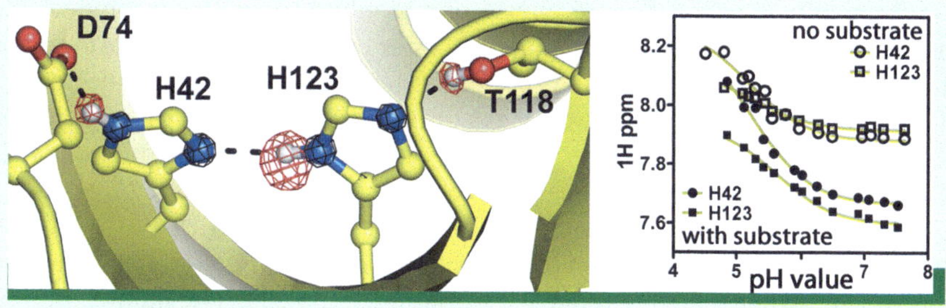

All's well that ends well. My meticulously metered titrations have eventually been published—Mueller et al. **2011**. Crystallographic proof for an extended hydrogen-bonding network in small prolyl isomerases. *J Am Chem Soc*. 133(50): 20096–9. ► https://doi.org/10.1021/ja2086195.

(Image credit: Modified reproduction with kind permission from Mueller et al. **2011**. *J Am Chem Soc*. 133(50): 20096–9. Copyright 2011 American Chemical Society.)

Context Local pH value and local pK_A values, they all are influenced by the environment. It almost seems to be magical when a tiny corner of the cell suddenly has a different pH value than its surroundings. Or when neighboring amino acids dramatically change the acid-base properties of a catalytic amino acid residue. But enough about buffers. Next, we will discuss what holds the world together in its innermost core.

1

1.3 The Air-Con is On, Keep Windows Closed—or—The Second Law of Thermodynamics. And Other Rules…

Life finds a way. A puzzling principle of life is the strangely sounding sentence: to reach equilibrium, one must be far, far away from equilibrium. This sentence only makes sense if we think of the two equilibria in different ways. The first equilibrium may be balanced fluxes or a **steady state**; the second, however, refers to a **thermodynamic equilibrium** in a closed system. ◘ Figure 1.3 shows visual input for both types of equilibria. How marvelous is it that we, "any living being", in this case, are able to achieve a fairly decent degree of stability with permanent flow-through of material and energy through our body. Some people call this stable state "normal" or the "normal state". However, the author of this book sees plenty of room for interpretation around what "normal" actually is.

Energy cannot be created out of nothing; energy cannot disappear into thin air. Instead, **energy can be converted from one form of energy into another form**. For us human beings, this flow of energy commonly starts with sunshine which is somehow stored by plants and animals and later comes to us in the form of food and drinks. With this input, our body generates high-energy molecules such as ATP in complex ways. These power packs of molecules in turn drive complex biosynthesis pathways and enable our muscles to carry out work, hopefully somewhat meaningful work. In the end, a pleasant feeling of warmth remains.

As the day lengthens, so the cold strengthens—keeping warm with ease is a comfy thing, especially on those cold days of spring. Even better, if that warmth comes to you, without having to have a workout by yourself. The humble bumblebee and brown fatty tissue, mainly found in newborns, are astounding in this

◘ **Fig. 1.3 Balance is not something you find; it is something you create**. Permanent energy conversion eventually results in some stability, a steady state. Such a steady state equilibrium was visualized using watering cans. On the contrary, a closed thermodynamic system might be compared to a room with air conditioning. Would we disconnect the room entirely, creating a perfectly closed system—perfect cooling would be guaranteed. We all know that this is a bit too good to be true. (Photo: Watering cans in the garden, Birmingham, JWM, 2014, sign, Bad Blankenburg, JWM, 2018; the quote in the title of this figure is attributed to Jana Kingsford)

regard. They are able to generate heat directly from chemical energy, without any muscle trembling. The biochemical tricks at work do vary between them. Sometimes they involve uncoupling proteins to bypass parts of the respiratory chain. Our very own brown fat tissue is being researched intensively. Wouldn't it be great if we could produce more of brown fat within the average overweight human adult—the excess fat would literally melt away.

And now for something completely different. Imagine, you are at a family function or a conference, where you admittedly didn't want to go to. You might feel bored, until... yes, until a clear broth of soup is placed in front of you. For our purpose, it doesn't matter if that broth is tasty or not. What matters is sufficient illumination, illustrative enough that you are able to see some oily droplets swimming around at the surface—and how they fuse with each other over time. This process is driven by perhaps the weirdest driving force in biochemistry. Actually, it's not a force at all. It is an effect, well, technically an indirect effect. It is the **hydrophobic effect**, we are speaking about. The hydrophobic effect is probably the **most important driver for the formation of biomolecules**.

As a start, the hydrophobic effect originates from a peculiarity of water—its very strong **hydrogen bonds**. With these, water can resolve charged particles really well. But with uncharged, non-polar molecules, water gets "frustrated" easily, not able to connect, without suitable docking sites for H-bonds on these particles. If water has to enclose hydrophobic matter still, an unfavored interface is created. At such interface, water molecules are forced to adopt a certain order—something that water doesn't particularly like, since ordered water particles represent an energetically (more precisely entropically) unfavorable state.

And now the hydrophobic effect comes into play: When two hydrophobic particles get close enough to each other, then water presses them together with a force that corresponds to the number of ordered water molecules that can be released. Energy is gained by reducing the number of energetically unfavorably arranged water molecules. The more water is released from order, the larger the indirect force the hydrophobic effect creates. There are other forces in biochemistry that remain unmentioned here.

The **volume-surface inequality** does not belong to these forces (◘ Fig. 1.4). Maybe more a principle of design that you should consider blaming mathematics for, instead of blaming biology for it. Please think of a cube with all edges with a length of 1 cm (well, it would work with inch as well; the unit doesn't really matter). This cube has a volume of 1 cm^3 and a surface area of 6 cm^2 Now, let's double the edge length —this could be quite a normal biological process: Some animal just grows to double its size. Back to the cube. It now has a volume of 8 cm^3 and a surface area of 24 cm^2. Now let's work out the volume-to-surface ratio: 1/6 = 0.17 versus 8/24 = 0.33, the larger cube has a volume-to-surface ratio TWICE as large as the small cube. This comparison may be regarded as the **volume-surface inequality**. Please note that increasing the difference in size makes this inequality increasingly unequal.

1

Fig. 1.4 **The volume-surface inequality**. Large and small organisms work in quite different ways. For a water strider (presumably *Gerris lacustris*), the surface tension of the water is vitally important. The water strider walks on the surface of water, they literally live on it. The insect finds its prey because of the waves the prey causes when it falls onto the water surface. Several orders of magnitude larger and heavier, the elephant might rather worry about the force of gravity. Mere falling down might already cause damage. Elephants also have to manage to get enough oxygen to all their tissues. Finally, cooling might be a problem-to get heat away from the core of the body. Have you ever thought of the elephant-sized ears as cooling devices? (Image credits: Elephant, South Africa, JWM, 2002; Water strider, Kassel, JWM, 2006, Cube, Birmingham, JWM 2022)

Construction The hydrophobic effect is a major driver in the formation of biochemical biomolecules. Various physical forces affect creatures of different sizes in non-uniform ways. To summarize this size inequality: Having the same overall body layout, once large and once small, does not mean that both body plans work equal, or even that they distantly work similar. So, it's not that easy to turn a mosquito into an elephant. Fun fact: "To turn a midge into an elephant" is another directly translated German proverb, put here to good scientific use. An English counterpart such as "to make a mountain out of a molehill", still is nice, but at least judged by the author, this one has less direct biological connotation.

Navigation

More about bumblebees and different types of adipose tissue in the meandering on metabolism in ▶ Sect. 6.5.

A 2017 study see glycerol-3-phosphate dehydrogenase at the forefront of the bumblebee auxiliary heater—Masson et al. **2017**. Mitochondrial glycerol 3-phosphate facilitates bumblebee pre-flight thermogenesis. *Scientific Reports*. 7(1): 13107. ▶ https://doi.org/10.1038/s41598-017-13454-5.

A fantastic early essay on how vastly different biology functions at different scales—Haldane JBS. **1926**. On being the right size. *Harper's magazine* 152: 424–427.

1.4 Could the Laws of Nature Change?

It is really impressive that water is so remarkable that it makes life possible. Computer simulations might enable us to study how it would be like if water had a different dielectric constant. Regardless, just like the elements of hydrogen and oxygen, water should be universal, something like a natural constant. That means, if there was water on a distant planet, then it should behave just like water on Earth.

In a way, our Earth is breathtakingly mediocre—it is best at being average. Our home planet is that far away from the sun that we are about in the middle of the habitable range. Our Earth has a rotating core of iron that generates a magnetic field, similar to a dynamo generator, that protects us from aggressive cosmic radiation. The Moon helps to stabilize the orientation of the Earth' axis, thus keeping the magnetic field stable. The Earth is just heavy enough not to have its atmosphere blown away by the interstellar wind, but light enough that we can walk upright on it. And many more seemingly lucky coincidences…

Astrobiologists describe this accumulation of favorable conditions as the **Goldilocks conditions**—everything is not too much and not too little, but **just right** (◘ Fig. 1.5). The environmental conditions on the neighboring planets Venus, Mars and Jupiter look worse by far. In the search for habitable planets, the search for an Earth 2.0, the Goldilocks conditions play an important role in evaluating if a planet—in principle—could support life as we know it.

The Goldilocks conditions apparently do not play a substantial role for the understanding of life that exists on our planet Earth, right now—they just have been there "forever". Well, nothing lasts forever. Not even the environmental conditions on our home planet. Let's imagine we completely fail to rescue the World from the imminent climate catastrophe. Humongous amounts of polar and glacier ice melt and sea levels rise worldwide, flooding large parts of our planet, including London in England and other densely populated areas. Welcome to water world, with conditions of life not too dissimilar to an exotic water planet. Hence it is good to know that the human body is adapted to life in water, remarkably well, already. Under those new watery conditions, our body plan would probably even evolve to some kind of aqua-men.

A study on the indigenous Bajau people, also known as Sea Nomads, of Southeast Asia sheds light on how quickly such an adaptation to water may have happened. For generations, the Bajau have been pearl divers—they have signifi-

■ **Fig. 1.5** **The Goldilocks conditions**. The Goldilocks conditions received their somewhat strange name from a British fairy tale from the nineteenth century. In it, the little girl Goldilocks visits the cottage of three bears, Papa Bear, Mama Bear, and Baby Bear, who handily are out for a walk. Goldilocks tastes from the three bowls of porridge and finds that the first is too hot. The second is too cold. Only the third is just right. She tests the three beds in the same way, one is too big, one too small, only the third is just right. And so on… Goldilocks seems to be popular not only in the search for life on other planets. A search for "Goldilocks[title]" on ▸ PubMed.gov yielded 395 hits [19th Feb 2025]. Enjoy browsing through all this literature. (Image: Wood stove, JWM, 2022)

cantly adapted to longer dives within a few generations (■ Fig. 1.6). Their molecular workup shows slightly altered thyroid function, that results in an enlarged spleen. And since the spleen stores oxygen-rich blood, a larger spleen helps to dive significantly longer. Somewhat sarcastically asking: If we can adapt to a life by the sea that easily, does anyone still fear the consequences of climate change?

The human species *Homo sapiens* is amazingly adaptable anyway. Much has been written about an N-terminal shortening in the CCR5 cytokine receptor pro-

Fig. 1.6 Humans can adapt to long dives. The Bajau people are sea nomads who regularly perform particularly long dives of 10 minutes, or more. Genomic sequencing revealed that changes in two different genes very likely are driving this adaptation to diving: The phosphodiesterase PDE10A, which cleaves the cyclic nucleotides cAMP and cGMP, has evolved in the Bajau people in particular. It directly influences the muscle activity at the spleen. PDE10A also influences the spleen indirectly, as it modulates the secretion of the thyroid hormone thyroxine. Both effects result in the spleen growing larger. And then there is genetic adaptation in the gene for the B2 receptor for the peptide hormone bradykinin; a signaling pathway involved in many things. In the context of the Bajau, this mutation optimizes the regulation of vasodilation during diving. Diving more regularly, may make your spleen larger, to store oxygen-rich blood. This is at least what has been observed in some kind of sea normads—Ilardo et al. **2018**. Physiological and Genetic Adaptations to Diving in Sea Nomads. *Cell.* 173(3): 569–580. e15. ► https://doi.org/10.1016/j.cell.2018.03.054. (Image credit: © Alex Photo ► stock.adobe.com)

tein, thus shrinking a putative docking site for the AIDS-causing virus HIV. Such N-terminally truncated CCR5 receptors might not have been something new. They are suspected to also have improved prognosis for medieval plague epidemics, and positive selection might have made such truncations spreading in western populations, in just a few generations. Same principle, but different receptor, the ACE-2 receptor for Covid-19 is expressed to varying degrees in different populations. These differences in ACE-2 expression could at least partially explain the varying severity of the course of the Corona infection observed in different parts of the human population.

The human body can adapt to changed conditions of the environment relatively quickly—maybe already over just a few weeks or months. Have you ever heard of **altitude training**? In this special form of training, athletes diligently run through their routines, but do so in the Alps, the Central Mexican Plateau, or else-

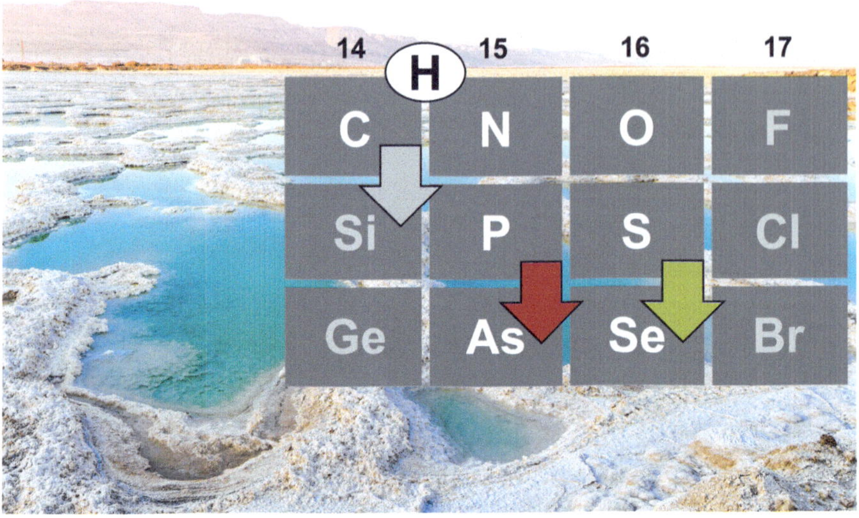

Fig. 1.7 Can life exist without the CHNOPS elements? Alternatives to certain CHNOPS elements should mostly be found within their respective groups of the periodic table of elements; most likely below them. A few thinkable replacements are marked with arrows and discussed in the main text. (Image credit: Dead Sea, © Olesya ► stock.adobe.com)

where high up there. The thin air up there contains less oxygen, stimulating the proportion of red blood cells to increase. Maybe an important advantage for an upcoming competition in the lowlands, as it doesn't hurt to have a little higher endurance than your competitors. Added bonus: Contrary to direct treatment with the "blood-making" hormone erythropoietin while exercising in the lowlands, altitude training is a legal form of performance enhancement. It's not the mountain we conquer, but ourselves.

Some elements of the periodic table are essential for life, with some being "more" essential than others. We call them the vital macro-elements: carbon, hydrogen, nitrogen, oxygen, phosphorus, and sulfur; abbreviated to CHNOPS. With the exception of hydrogen, all these elements can be found in groups 14 to 16 of the periodic table. Within Fig. 1.7 you might even read "CHNOPS" with squinted eyes, just give it a try.

Could life exist without using one of the CHNOPS macro-elements? Counter question: Can you write a longer text without using a frequently occurring letter of the alphabet, the "R" for example? This form of literature is also regarded as a lipogram. A great lipogram is "The silence that followed a strong thunderstorm" by the German Baroque poet Barthold Heinrich Brockes. The stRong thundeRstoRm is descRibed with numeRous RRRolling R's—the long text about the silence that follows instead does not contain a single copy of that growling consonant at all.

Back to the CHNOPS question. We have arrived at hypothetical biochemistry—that sort of biochemistry that could exist on distant planets or, more likely, the biochemistry of some yet undiscovered, highly exotic microorganism on our Earth. You may want to think of plastic-degrading bacteria that recently were covered in the media. The most likely changes can be seen in Fig. 1.7. The exchange of sulfur by selenium is everyday business in our cells' biology—a fair few special redox proteins have selenocysteine incorporated in their catalytic center, as a kind of super-cysteine. Silicium is an element grouped together with carbon in the periodic table, listed directly below that versatile element. So, could silicium be used instead of carbon at some point or another? We don't know. Finally, could phosphorus maybe be replaced by arsenic? We do know a little more about this one—a dedicated meandering text box provides more information about life without phosphate.

Meandering: Could Life Exist Without Using one of the CHNOPS Elements?

It was big headlines, back in 2011. A research team from NASA claimed to have found a bacterium that uses the otherwise toxic arsenate—instead of phosphate—for life. From December 2010 onwards, there were press releases and various press conferences, followed by the publication of the actual research data in the journal Science in June 2011. The NASA team, led by Dr. Felisa Wolfe-Simon, really were pushing it; exceeding "normal" amounts of science communication and advertisement, "to a certain degree". The backlash beat-up—in the form of appropriate scientific criticism—followed quickly and heavily. It doesn't happen very often that a publication is published accompanied by eleven (!!!) contradicting comments. Regardless of whether the claim by NASA was true or not, the paper accumulated a lot of citations quickly. It seems that the saying "any publicity is good publicity" does apply to science as well.

Found in Mono Lake in California, the newly isolated bacterial strain could grow under very high, otherwise certainly toxic concentrations of arsenate. Mono Lake is an intensely salty and alkaline lake—you could probably make decent German pretzels or lye rolls in there, as these delicacies are traditionally dipped in alkaline solutions before baking. The new bacterial strain thrived, seemingly in the absence of phosphate. So, what's the trick of this bug? Somehow the authors of the study were able to convey the impression that this microorganism uses arsenic as a true substitute for phosphorus. Big news for the CHNOPS elements, from now on without phosphorous; is it really CHNOS then, without the P?

Between 2010 and 2012, these findings caused a lot of debate. You can still read some of the controversy, if you followed the hashtag #ArsenicLife. Some praised that the internet made it possible to praise the scientific method itself as praiseworthy. Others thought that it might not suit science too well, if a potentially important new development was fed to sensationalist reporting at a too early stage.

The bone of contention—Wolfe-Simon et al. **2011**. A bacterium that can grow by using arsenic instead of phosphorus. *Science*. 332(6034): 1163–6. ▶ https://doi.org/10.1126/science.1197258.

After a lot of debate, the fairy tale of the arsenic-loving bacterium was irredeemably axed by two papers in 2012—All nicely introduced here by Rosen RJ, **2012**. The Case (Study) of Arsenic Life: How the Internet Can Make Science Better. *The*

1

Atlantic. ▶ https://www.theatlantic.com/technology/archive/2012/07/the-case-study-of-arsenic-life-how-the-internet-can-make-science-better/259581/ [Accessed: February 19, 2025].

These **two** studies showed that the new bacterial isolate can tolerate arsenic very well, but that nothing, really NOTHING of it is built into its own DNA—Erb et al. **2012**. GFAJ-1 is an arsenate-resistant, phosphate-dependent organism. *Science*. 337(6093): 467–70. ▶ https://doi.org/10.1126/science.1218455—**and**—Reaves et al. **2012**. Absence of detectable arsenate in DNA from arsenate-grown GFAJ-1 cells. *Science*. 337(6093):470–3. ▶ https://doi.org/10.1126/science.1219861.

Common Knowledge It certainly is good to put old dogmas to a test, every now and then. For now however, it remains a scientific fact that most biochemistry is accomplished WITH the CHNOPS elements, and not without them. Regardless, extraordinary chemistry is mastered by amazing cofactors, as we will discuss later.

Further Reading

A lot has been published about Corona. Here is a relatively early publication that looks at the different course of a Corona infection relative to ethnic origin—Raisi-Estabragh et al. **2020**. Greater risk of severe COVID-19 in Black, Asian and Minority Ethnic populations is not explained by cardiometabolic, socioeconomic or behavioural factors, or by 25(OH)-vitamin D status: study of 1326 cases from the UK Biobank. *J Public Health (Oxf)*. 42(3): 451–60. ▶ https://doi.org/10.1093/pubmed/fdaa095.

We will discuss how selenium sneaks into proteins, when looking at the peculiarities of the genetic code in ▶ Sect. 3.6.

1.5 Oxidation Numbers and Some Stereochemistry Really Help You to Better Understand the Biochemistry of Life

Until now, we have discussed water, thermodynamics, and the essential macro-elements CHNOPS. What else is needed to better understand natural compounds? **Oxidation numbers** would come in handily here. Let's nicely count electrons according to the "*oil-rig*" mnemonic. However, please don't think of drilling islands out at sea (also called oil-rigs); instead let's pretend that *oil-rig* stands for "*Oxidation Is Loss, Reduction Is Gain*" (■ Fig. 1.8). Plainly, oxidation numbers become smaller in an oxidation reaction, and get larger in a reduction. A lot of biochemistry is centered around reducing carbon atoms and oxidizing them again. Very likely, high school graduates get confronted with this eminent biochemical equation:

$$C_6H_{12}O_6 + 6O_2 \rightleftarrows 6CO_2 + 6H_2O$$

Oxidation numbers help to determine where the interesting stuff happens in the middle of a lot of other biochemical stuff—it is the reduction/oxidation reactions

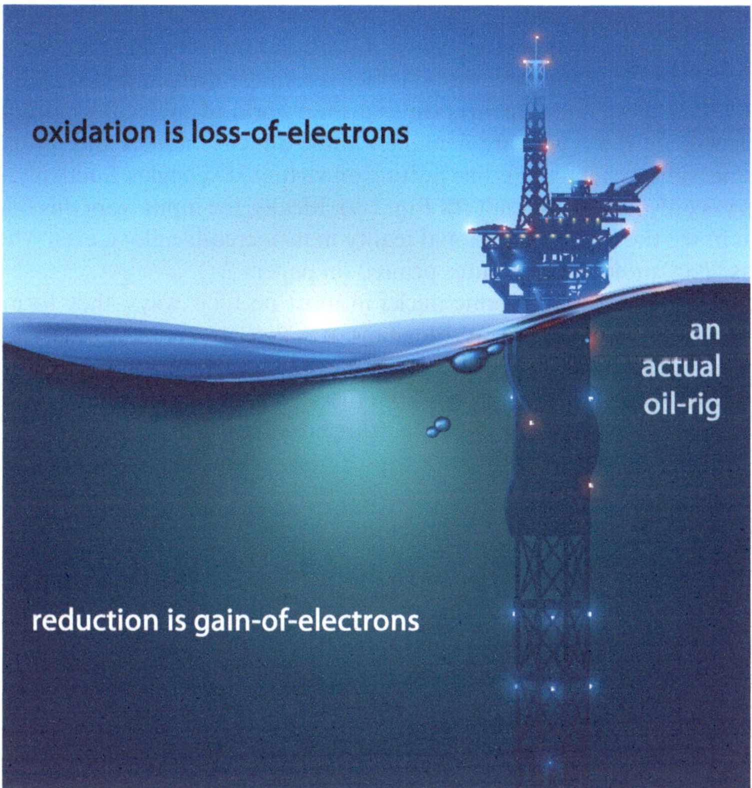

Fig. 1.8 **An actual oil-rig may help to understand oxidation and reduction reactions**. Feel free to "walk" within this analogy as long as you wish. In the figure, I have put oxidations into air and reductions into water; guess where more oxygen would be available. (Photo: Oil-rig, © AndSus ► stock.adobe.com; adapted by JWM)

(Red-Ox, or simply redox), where electrons are withdrawn from somewhere and then added somewhere else.

Oxidation numbers reduce complexity. They simplify the charge distribution in molecules, assuming that these compounds were monoatomic ions instead. Hydrogen and oxygen are listed in glucose as H^{+I} and O^{-II}—thereby giving carbon an oxidation number of zero "0", to make the entire molecule electrically balanced. Somewhat advanced rules say that bonds between two atoms of the same element are shared equally. Thus, the oxygen in hydrogen peroxide gets a "−1" as an oxidation number, which is a notable exception.

$$C^0{}_6H^{+I}{}_{12}O^{-II}{}_6 + 6O^0{}_2 \rightleftarrows 6C^{+IV}O^{-II}{}_2 + 6H^{+I}{}_2O^{-II}$$

Circle A large part of life on Earth deals with oxidizing and reducing carbon between its oxidation states −IV and +IV. On the go, pure oxygen is used up and newly made again. All these processes produce CO_2 and H_2O, puffy and humid air. Oxidation numbers are moderately difficult, but are quite handy, due to their great predictive power. With oxidation numbers, we can confidently examine the many redox reactions that occur in the cell at any time. We will discuss some of them in this book.

...and then there is the realization that **the "world of molecules" is not a flat disc**. Instead, its molecules are genuinely three-dimensional! A tiny infusion of three-dimensional chemistry is like putting on virtual 3D goggles. Suddenly, **stereochemistry** is only half as difficult (◘ Fig. 1.9). Ideally, the input from this book will help you to see the three-dimensional reality in its incredible glory, even when presented as "flat" molecule structures printed on paper.

Electrons orbit around atomic nuclei in most peculiar ways; they form oddly shaped clouds of electron density—for fans: it is these patterns of electron probability, we call orbitals. Orbitals have fancy names, such as "s", "p" or "d" orbitals. In their outer electron hull, elements from the second and third period of the periodic table have a spherical s-orbital and three dumbbell-shaped p-orbitals, which are at a 90° angle to each other. If these p-orbitals would be involved in molecular bonds directly, their binding geometries should primarily be right-angled bonds—everything should be nicely orthogonal.

◘ **Fig. 1.9 Acetate—a tetrahedron and a small collection of trigonal bipyramids. Left:** The CH_3 methyl group of acetate shows up beautifully tetrahedral. It is juxtaposed with a massively large sculpture in Bottrop, Germany, called the tetrahedron. The C-atom would sit in the middle of the tetrahedron—somewhere below the upper observation deck. That steel tetrahedron in Germany comes with a side length of 60 meters; you might be able to spot a few people in the picture to serve as a proxy for size. **Right:** The carboxyl group of acetate is flat—a planar collection of three small trigonal bipyramids, two for the two oxygen atoms and one for the carbon atom. One of these double pyramids is represented as structural fruit. The carbon would be the grape; the two oxygen atoms are represented by the two blueberries to the right; the two lobes of the p-orbital are represented by delicious plum tomatoes. **Middle:** The single bond that connects both parts rotates freely. And by the way, a lot of daily-life-objects could be useful to study stereochemistry. If you don't want to give fruits and veg a try, why not try licorice or so? (Image credit: Tetrahedron, JWM, 2006; Structural fruit, JWM, 2021)

At least carbon, nitrogen, and oxygen from the second period of the periodic table do something different, however. For their hydrogen compounds methane (CH_4), ammonia (NH_3) and water (H_2O), bond angles of 109.5°, 106.7° and 104.5° have been measured experimentally. It can be read in chemistry books, printed or online. That's the way it is!

To explain these weird shapes, we can draw on two theories. One is the **VSEPR model**, the "valence shell electron pair repulsion" model for electrons in the outermost electron shell. This model assumes that an electron pair in the outer electron shell doesn't care whether it comes from an s-, a p-, or any other form of orbital. Instead, being doubly negatively charged, it tries to keep away from the other electron pairs, as far as possible. Distantly similar to a dinner party during Corona—*a socially distanced electron pair*. If all electron pairs have identical bonding partners, as in methane (CH_4), then we end up with the perfect **tetrahedral bond angle** of 109.5°. Free electron pairs, however, need a bit more space, understandably so, as they are not kept in check by bonding. The roaming free electron pairs compress electron pairs in covalent bonds a little; this is what makes water molecules to have a somewhat smaller bond angle.

A different view on the same thing is that the elements C, N, and O, and maybe more elements, can mix their round s- and dumbbell-shaped p-orbitals of their outer hull. This mixing, also known as **hybridization**, may involve one s-orbital and three p-orbitals, making this mixing an sp^3-hybridization with four identical orbitals, to engage in four single bonds, all extending from the center of a tetrahedron—immediately we are back at the carbon in methane with a bond angle of 109.5°. Double bonds look quite different; plausibly they can be explained by an sp^2-hybridization. A trigonal bipyramid is created. It is this binding geometry that can be better understood via hybridization than relying on the VSEPR model. Whatever is used for explanation, the different stereochemistry of the two carbon atoms in the acetic acid residue, the acetate ion CH_3-COO^- can be clearly seen in ▫ Fig. 1.9.

The element carbon in its pure form already shows these two types of bonding. **Graphite** is a very pure version of coal, in which the carbon is sp^2-hybridized—the material conducts electrical current because the non-hybridized p-orbitals overlap with each other. The other form of carbon has all atoms nicely sp^3-hybridized—it is called a **diamond**.

> "A diamond is a chunk of coal that did well under pressure", a quote attributed to Henry Kissinger.

> The author of this book came up with this one: "Perhaps Time's definition of coal is the diamond."

Regardless, in a diamond, the more "social" form of carbon, each atom forms strong covalent single bonds with all neighboring atoms, making diamonds one of the hardest materials we know. However, there are no overlapping p-orbitals—that's why diamonds cannot conduct electrical current. Regardless, diamonds usually are regarded to be much prettier than the graphite core of an ordinary pencil.

Compelling Within the 3D world of biochemical molecules, the elements C, N, and O show bonding geometries of almost perfect tetrahedra, as long as it all is about single bonds. When it comes to double bonds, however, these elements all of a sudden switch to trigonal bipyramids. Then, all bonding partners are flat in one plane. Perplexingly perpendicular to this plane are the p-orbitals; they confer the true character of a double bond. Practically, this means that a **single bond** involving at least one carbon **is rotating freely** around the axis of the bond. A **double bond**, however, **is not free**, it doesn't rotate, but is locked in its rotation.

- **Navigation**

More about diamonds and molecules that look like diamonds, in ▶ Sect. 2.5.

1.6 What is Chirality: Experience it First-Hand…

On the one hand, oxidation numbers and three-dimensional chemistry already provide a good basis for understanding natural chemical compounds. *On the other hand*, a decisive ingredient is still missing, and this is **chirality**. Please take a look at your hands—they behave like image and mirror image to each other. No matter how much you twist or turn them, it is impossible to perfectly overlap both hands on top of each other. Handily, the word *chirality* itself means handedness.

Carbon can bind up to four other atoms, and if all of these are different, this very carbon becomes **chiral or handed**. Suddenly there are two versions of said carbon that cannot be overlaid easily—neither by rotating or by shifting. Such a pair of mirrored molecules is called enantiomers. Shining polarized light on quite many pairs of enantiomers may reveal a funny difference between the two molecules, one might bend light to the left, while the other one does the opposite, it turns light to the right. Therefore, people have come up naming these pairs of molecules **L- and D-enantiomers**, in remembrance to the Latin words *laevus* for left and *dexter* for right.

Some pairs of enantiomers also differ in their biochemical properties. An example is the different taste of the amino acids L- and D-valine. You can see this pair of L- and D-amino acids in ◘ Fig. 1.10; accompanied by the actor Samuel Jackson. A chemically chiral meme in social media was based on Mr. Jacksons middle name Leroy. The figure is enriched by a very special gift for Valine-tine's Day.

L- and D-valine can both be bought as white-yellow-ish powders. Both share the same molar mass, the same melting point, the same solubility in water, and so on. Chemically speaking, they are identical, nearly. Chemists will struggle to tell them apart. Biologically, however, these molecules differ greatly from each other. For our tongue, L- and D-valine have a completely different taste (of course only theoretically, since we would never taste chemicals in the lab): L-valine has a bitter taste, while D-valine comes along rather sweet. More importantly, the ribosomes of any living being would use L-valine, exclusively; ribosomes make proteins exclusively of L-amino acids, never ever from D-amino acids.

An exception to the L-amino-acids-only rule are extraordinary anti-bacterial peptides, made by some bacteria, using not their ribosomes, but some very special peptide tomboy knitting devices, so-called non-ribosomal peptide synthetases.

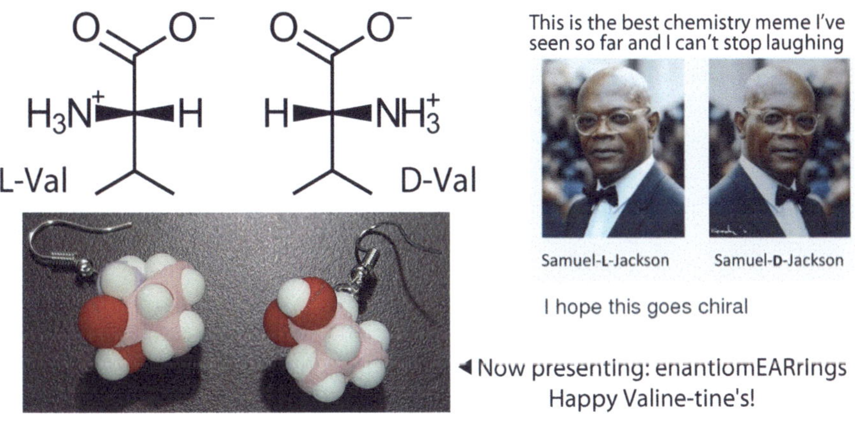

Fig. 1.10 A special pair of amino acids, an early chemical meme, and a recent addition. Top left: L- and D-valine are a pair of enantiomers that behave like image and mirror image. Only L-valine is incorporated into proteins within the cell—our ribosomes will never "touch" D-valine. **Top right:** Samuel-L-Jackson and his mirror image Samuel-D-Jackson are a chemical meme, that went viral, or chiral, some years ago. The two Jacksons certainly inspired many variations and interpretations. A modern classic that now has found its place in a textbook. **Bottom:** Now introducing enantiomers as earrings: valine enantiomEARrings. And if you think, it cannot get any better: Wearing them on February 14th, makes them ever so much Valine-tine's-ier. Thinking of loved ones, biochemically speaking. (Image credit: Tweet from @Salonium_34, 9.7.2017, on what then was called Twitter; Reposted by JWM on BluSky; Valentine earrings, re-used with permission from Leah Krevitt)

These assembly lines produce complex chemical defense substances, with in-build D-amino acids and other tricks to enhance chemical stability.

Complicated Chirality Unlike quite many ordinary organic reactions, the vast majority of enzymatically controlled conversions run under stringent stereochemical control—most enzymes work with astonishing stereoselectivity. Hence, biochemical building blocks often exist as image and mirror image; and these variants are referred to as pairs of enantiomers. Biochemically speaking, it is highly relevant which of them should be used. In the cell, there tends to be an absolute separation of enantiomers. Only one of them is incorporated into biomolecules, such as L-amino acids in proteins and D-sugars in RNA and DNA.

Navigation

In ► Sect. 2.7, we will explore why we use these biochemical building blocks to build biomolecules, but no others.

Bumblebees and the brown adipose tissue of newborns are incredibly fascinating. Because they have found various ways to decouple the respiratory chain from ATP synthesis, they can directly convert chemical energy into heat, without muscle trembling. A meandering within ► Chapter 6 deals with bumblebees.

The LEGO Construction Kit of Life

Contents

© The Author(s), under exclusive license to Springer-Verlag GmbH, DE, part of Springer Nature 2026
J. W. Mueller, *Ultimately Understanding Biochemistry*, https://doi.org/10.1007/978-3-662-71889-6_2

Please think of just a handful of atoms of the elements carbon, hydrogen, nitrogen, and oxygen, and all the many ways to combine them with each other; it is surprisingly many possibilities already. Adding only a few more atoms, creates MANY more possibilities. Even for quite a few atoms of C, H, N, and O, the theoretically possible chemical space offers an enormous number of combinations. Therefore, categorization of compounds is a must, unfortunately. Probably the best way of giving you a feeling of what is possible here is via calculating the maximum possible number of combinations, the "chemical space". This is attempted via the calculation on **napkin III**. Possibilities explode when the elements sulfur and phosphorus are added. Collectively, these six elements are called the CHNOPS elements; they make up the vast majority of living matter.

Napkin III—What is Possible Within the Chemical Space

C, N, and O atoms are estimated with their "normal" valency of 4, 3, and 2 respectively. H with a valency of 1 is not counted and will be added later. Steric hindrance is ignored at this stage. $C_2N_1O_2H_x$ = would therefore be $4^2 \times 3^1 \times 2^2$ or already 192 possibilities to arrange these few atoms alone, speaking of non-hydrogen atoms ones. Probably only ONE of these 192 possibilities is the amino acid glycine, the **smallest** amino acid that is built into proteins.

The name glycine is derived from the sweet taste of pure glycine and the Greek word *glykós* for sweet. Therefore, **the accompanying figure** shows the chemical structure of glycine paired with some crystal-clear sugar to illustrate "sweetness". Admittedly, not the best pairing in this book, as I used a picture of sucrose, also known as saccharose. I still hope you regard this picture as useful.

As we used some powdery sucrose on this **napkin III**, for comparison, we might want to re-do the chemical space calculation for sucrose's overall formula—$C_{12}H_{22}O_{11}$ (one oxygen atom less than carbon, due to the connection of two sugar units). This empirical formula translates to $4^{12} \times 2^{11}$ for the calculation of all chemical possibilities. This is 34′359′738′368 possibilities already, more than 34 billion possibilities to stick together those atoms, that normally make up plain table sugar.

(Image credit: Crystal sugar, JWM, 2022)

Considering the vast chemical complexity that already a di-saccharide compound might offer, it would be good to divide the chemical diversity into meaningful units. But what is meaningful? **What do I gain from knowing**, for example, **that glucose is a reducing sugar?** Actually, quite a bit. This labelling contains a lot of information, some of it is revealed in Fig. 2.1. Compound classification might

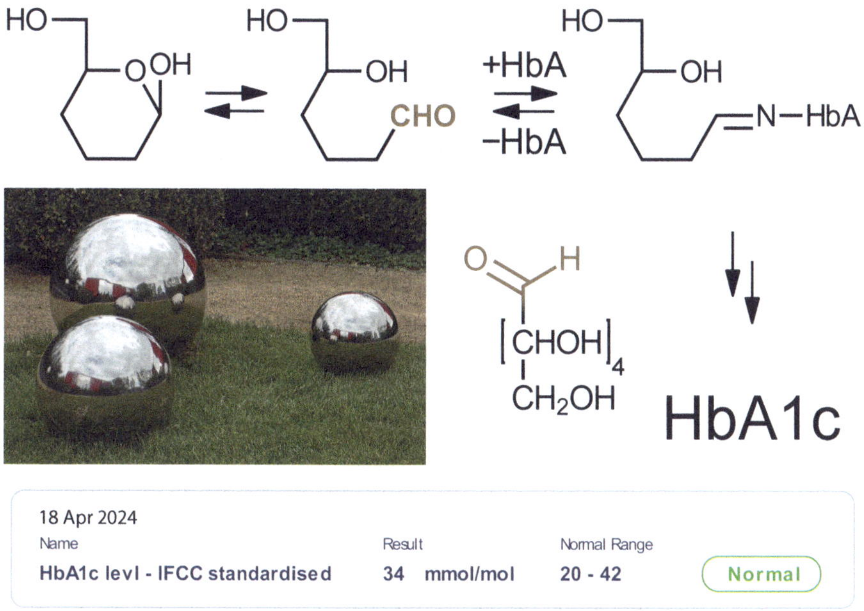

18 Apr 2024

Name	Result	Normal Range	
HbA1c levl - IFCC standardised	34 mmol/mol	20 - 42	Normal

Fig. 2.1 Glucose—part one—a reducing sugar. Glucose—or grape sugar—exists in two major forms in watery solutions. There is the ring form of glucose, and an open form of this sugar with a clearly recognizable aldehyde group (highlighted in ochre). The two forms interchange with each other; they are in equilibrium. The aldehyde group in the open form makes a glucose solution testing positive in the silver mirror test, where soluble silver ions Ag^+ are reduced to metallic silver—one could use this test to make beautifully silvered Christmas tree decorations. The very same aldehyde group (again highlighted in ochre) makes glucose spontaneously reacting with free amino groups of hemoglobin within blood. And there are plenty of amino groups in hemoglobin. Usually, the free N-terminal end of the protein bonds with glucose; but internal lysine side chains are reactive as well. First, the aldehyde of glucose and an amino group form a Schiff base R-C(H)=N-R', which then stabilizes itself in some complicated ways. For fans: This chemical rearrangement is called an Amadori rearrangement. Ta-dah, glycated hemoglobin, also called HbA1c, is formed. A person's HbA1c value is also called their long-term blood sugar, because this value reports the average blood sugar level of the last 8–12 weeks, which is the approximate lifespan of human red blood cells. (Photo information: Polished metal balls, Wernigerode, Germany, JWM, 2006; HbA1c test result, JWM, 2024)

be distantly compared to the classification of cars into diesel, petrol and electric, or into small cars and SUVs, and so on. Some of these classes are meaningful (petrol vs electric), some are more aesthetic (gray cars vs non-gray cars). Here we want to classify according to biochemical substance classes. We start with sugars (glucose et al.) and then discuss the special features of proteins, fats and oils, nucleic acids, and finally turn to isoprenoids.

More extensive substance classifications could be done according to biochemically relevant features, for example molecular weight, polarity or presence and absence of certain functional groups. You can classify natural substances according to their origin from within nature, such as endogenous (body's own) or exogenous (foreign to the body). Why not class biological molecules according to their function or effect, such as antibiotics, enzymes, hormones or vitamins. Some biochemical classifications are more meaningful or useful than others; very similarly to classifying cars. All of them, hopefully, might help to manage complexity.

The central biochemical pathways involve only relatively few substances; contrasted to the humongously large variety of substances in organic chemistry and beyond. Biochemical substances within this core set often have historical names, which might even be different in different languages. α-Ketopropionic acid, for example, is "pyruvic acid" in English, while German biochemists affectionally call this molecule *brenztraubensäure*. The α letter points to another peculiarity of biochemical nomenclature: "We biochemists" number carbon atoms in carboxylic acids differently to chemists. We do not count the carbon atom of the carboxyl group itself; instead, the second C atom traditionally is called the α-C atom, and the third one is β. This results in such beautiful terms like "α-amino acids" (discussed further in ▶ Sect. 2.2) and "β-oxidation" of fatty acids (in ▶ Sect. 7.5). Good old days.

Core Set In biochemistry, we only deal with a small, but still considerable selection of formulas, substances, and metabolites. These usually have a long history—both evolutionary and in the history of science. These substances almost always carry historically grown trivial names. Learning biochemical facts is often perceived as complicated. Please try to add this combination of facts and formulas to the mix, to make it more relatable and a bit easier to learn.

■ **Navigation**

We will compare some of the amino acids that make up proteins, to delicious things. The link between tryptophan and chocolate is discussed in ◘ Fig. 2.5.

2.1 Sugars—Life has Never been Sweeter

Everyone knows that sugar is sweet. Wait a minute. From day-to-day experience, we actually only know that exactly two sugars are sweet. And these are the sucrose crystal sugar and the grape sugar glucose. We will start with some general remarks about sugars, their empirical formula, and the classification of sugars into reducing and non-reducing sugars. Then we will introduce some types of bonds commonly found

Fig. 2.2 **Carbohydrates are carbon hydrates**. A bit of charcoal and a sip of water, that and not much else, are the chemical ingredients for sugars. From their chemical formula, sugars resemble formaldehyde, or its polymerized form, para-formaldehyde. Isn't this sweet? (Photo information: Charcoal with water, JWM, 2022)

in sugars… A meandering will introduce you to one of the most expensive sugars on the planet; I'm talking about heparin. Finally, we will stretch the concept of sugars a bit by dealing with dietary fibers—certainly sugars, but definitively not sweet—and with sweeteners, which can be very, very sweet, but usually are no sugars.

Most sugars follow this **chemical formula** $\mathbf{(CH_2O)_n}$ reasonably well. Literally, sugars are exactly what their other name "carbohydrates" means. Sugars are molecules of hydrated carbon atoms, "carbon hydrates". For each carbon atom about one water molecule is bound, which brings us back to the formula $\mathbf{(C + H_2O)_n}$ (Fig. 2.2). Chemically, sugars resemble polymerized formaldehyde, so-called para-formaldehyde. Yummy!

Burning, or oxidizing, an aldehyde to a carboxylic acid, generates quite a lot of energy, as the carbon rises from oxidation number 0 to oxidation number +II; even if this process is dwarfed, in terms of energy, by the burning of fats. Contrary to fats though, sugars are outspokenly polar and soluble in water—with all their hydroxyl groups—making them readily available in the cytoplasm of a cell. This love for water makes storing sugars a difficult metabolic endeavor. A biochemical solution is the polysaccharide glycogen. Nicely branched, and branched again, **glycogen beads** are a great **storage form of glucose**.

Sugars can be split into the categories "reducing" and "non-reducing". **Reducing sugars** contain an aldehyde group; the chemist could call it a terminal keto group. For oxidizing this very aldehyde group into a carboxylic acid, this type of sugar can reduce something else in turn. Please look at Fig. 2.1 again, thinking of beautiful Christmas decorations that could be made using reducing sugars to silver glassware. Our grape sugar glucose sits within this category, even if you don't see the aldehyde group at first glance. On the contrary, **non-reducing sugars**, such as fructose, contain a keto group in the middle of the molecule that cannot be oxidized that easily.

Coming back to the question from the beginning of this chapter: **Why should I care about knowing that glucose is a reducing sugar?** Well, as a start, you should not use glucose, also known as dextrose, for preserving fruit, for making jam, marmalade, or chutney. This is because the aldehyde group would unsavorily react with amino groups on all sorts of biomolecules—resulting in changes of color, flavor, and maybe even texture. A much better choice for preserving food is sucrose.

The same reaction that renders glucose useless for making jams and marmalades, also happens in blood, where glucose binds and chemically glycates many proteins, one of them being our main oxygen-carrier, the protein hemoglobin. When blood sugar levels are elevated for an extended period of time, glucose can also damage blood vessels; hence, the unpleasant, but understandable, side effects of *Diabetes mellitus* are eye and vascular diseases.

Glycated hemoglobin, also known as HbA1c, may be bad for our vessels, it does however have an analytical advantage (by all means, please look at ◘ Fig. 2.1, yet another time). The glycating reaction between glucose and hemoglobin is spontaneous, it doesn't require an enzyme's action. This reaction also is irreversible; meaning, once the sugar is attached to hemoglobin, there is no way back. As a result, the measurement of the amount of HbA1c in our blood, can also be called long-term blood sugar. The HbA1c level corresponds to the average blood sugar of the last couple of weeks. All of this solely is based on the fact that glucose is a reducing sugar.

There are quite a few other ways to categorize the many sugars found in Nature. Counting the carbon atoms of your sugar molecules, you can observe trioses, tetroses, pentoses, hexoses, and heptoses in biochemistry; meaning sugars with three, four, five, six, and seven carbons. Pentoses and hexoses are somewhat the norm; all other sugars are quite exotic. You can categorize sugars by the ring sizes they form, and of course, whether they are L- or D-sugars. In biochemistry, D-sugars are used almost exclusively. Complex sugars can also be divided according to the number of sugar units into oligo-, and polysaccharides, contrasted to the more simple mono-, and di-saccharide forms.

Sugars are rich in hydroxyl groups, and this makes them generally highly soluble in water. With many functional groups to choose from, linking sugars to chains results in a high structural diversity, and diversity increases the longer the chains of sugars get. We call such chains polysaccharides or glycans.

Connecting Sugars Sugars are linked to each other by rather special bonds, and the "**glycosidic bond**" is probably the most special type among these bonds. The glycosidic bond forms between an OH group of a sugar on one side and an OH or an amino group of something else on the other side. But the sugar's OH group is not an ordinary OH one. It was actually a keto group or an aldehyde beforehand. Chemists like to call this stealth carbonyl group a half-acetal—an aldehyde-in-disguise, after forming a ring structure. Should the other reaction partner have been a nucleobase, we would now have a nucleoside (see the dedicated ► Sect. 2.4 on nucleic acids).

A truly clever sugar is crystal sugar, also known as **sucrose**. In sucrose, glucose and fructose are linked together via their reactive OH groups (those weird OH groups from above). This double **glycosidic bond** effectively hides these reactive groups. The result is an extremely robust sugar, which for sure is very well suited for jam-making or baking. Sucrose is inert; sucrose will not do any harm to your boiled fruit, chemically speaking. Just as a side note: Sir Walter Norman Haworth worked out the structure of sucrose in his time at the University of Birmingham, England. In addition, Haworth made such a great contribution to sugars that one of the ways we nowadays depict sugars has been named after him—the Hayworth projection.

Other Types of Sugary Bonds An **ester bond** is a connection of a "normal" OH group of a sugar molecule with whatever acid; acids normally would be phosphoric acid or a carboxylic acid, but could also include sulfuryc acid. Glucose-6-phosphate (Glc-6-P) is a good example for a possible product of such linkage. Such a "simple" ester bond is quite easily made undone by hydrolytic enzymes; similar to the glycosidic bonds, mentioned above. Somewhat unusual are deoxy sugars; those are sugar units where an oxygen atom is missing at a certain carbon. We will discuss deoxy sugars in the context of DNA in ▶ Chapter 3. Chemically thinking, one would also expect **ether bonds** in glycans; these could be made between two "normal" alcohol groups from various sugars. However, such an ether is not hydrolyzed easily. It is close to never being observed in biochemistry, with the exception of some extremophiles, maybe. Probably, this type of ether bond is chemically too stable to play a significant role in biochemical daily life.

We know that sugars can form quite a few different types of covalent bonds, often mixed. Very often, more than one bond is formed. This richness of bonds makes glycans biochemically so diverse. At the same time, however, sugars used to be a nightmare for the synthetic chemist; they are still challenging to this day, even though that field has seen quite some progress recently. This probably explains why heparin is such an expensive kind of sugar, explained in more detail in the next **Meandering**.

Meandering: A Very Expensive Sugar—Heparin

Heparin is a special kind of sugar, more precisely, a rather large sugar molecule—a polysaccharide. To be more precise, heparin is a group of complex sugars with built-in amino groups… And these sugars are covered all over with sulfate groups just like a cake is covered with icing. However, heparin is not used to sweeten tea...

Heparin is a very effective **inhibitor of blood clotting** and is used worldwide as such. The global sales of heparin products amounted to about 7.0 billion US$ in 2020. We are getting older, we injure ourselves more often; we want to protect ourselves from thrombosis, stroke & heart attack—that's why the use of heparin is more likely to increase than decrease.

A special motif of five sugar molecules, depicted in the accompanying **Figure**, binds to an inhibitor of blood clotting. This motif contains plenty of sulfate—eight sulfate units in total—shown in bold orange; while GlcA stands for β-D-glucuronic acid and IdoA for α-L-iduronic acid, the other sugars are probably more for fans. The sulfo-"coat" might be distantly similar to sugar icing in.

2

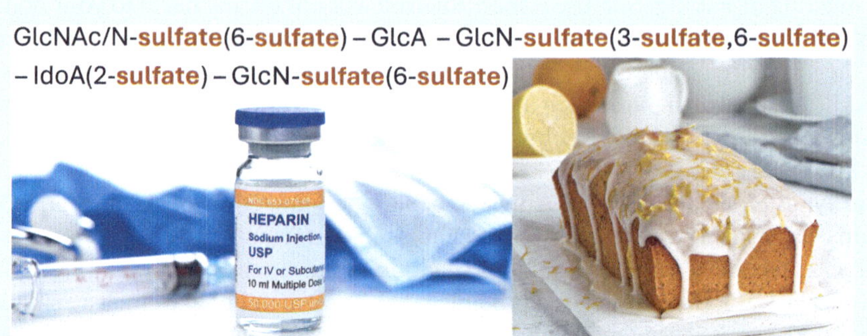

We could compare heparin sales to the entire global sugar market. Here, sugar refers to the crystalline sugar sucrose only. With 187.3 million tons of sugar production in the 2019/2020 season at about $0.24 US/kg, we arrive at $44.9 billion US, only about seven times as much as heparin... It is quite difficult to estimate the weight for the globally used heparin, as it is usually administered in physiologically effective units, and not in grams or milligrams. However, it is certain that it is a significantly smaller pile than that mountain of crystalline sugar.

(Image credits: Heparin, © Sherry Young ▸ stock.adobe.com; Cake, © Nata Bene ▸ stock.adobe.com)

Sweeter Than Sweet—Part I Automatically, we might expect all sugars to be sweet and that everything sweet is sugar. Think of good old sucrose, the normal table and crystalline sugar. Could it possibly get any sweeter? A rather traditional path to more sweetness is the **production of inverted sugar syrup**. Here, sucrose is split into its components, glucose and fructose; and this could be done either chemically or enzymatically. Bees do the same thing—they split sucrose into its components during honey production. Fun fact: Sucrose can rotate certain types of rays of light; and the strange name "inverted sugar" comes from the fact that this reaction results in a reversal of that rotation of light. Because the released fructose is perceived as very sweet, the resulting syrup is a sweetener, widely used in the food industry until today.

Not All Sugars Are Sweet Sugars that are not sweet at all are cellulose and other plant fibers, which we sometimes refer to as **dietary fibers**. They are indigestible components of plant foods that aid digestion. Dietary fibers bind water, hence generate volume and thus contribute to a better feeling of satiety. They also help digestion mechanically by lubricating the ingested food mass for a better journey through our intestines. On top of all that, the one or the other of our intestinal bacteria really like dietary fibers as its nourishment. Dietary fibers are contained in numerous types of vegetables. Without being a nutrition expert, I recommend eating some greens from time to time.

Sweeter Than Sweet—Part II There is more about sweetness than sugars, as we know quite a few substances that have a much stronger sweetening power than sucrose. These **sugar substitutes** are chemically diverse. I mention just three substances as examples: the natural product stevia, the dipeptide aspartame, and sugar alcohols, which are used widely in chewing gum (◘ **Fig. 2.3**). **Sugar alcohols** like mannitol or sorbitol can be produced industrially by hydrogenating sugars. They are not the sweetest things in the world; often only just reaching the sweetness of crystalline sugar. However, sugar alcohols contain almost no digestible calories. Consequently, not just our guts are starved, but also our caries-forming bacteria in the mouth. Chewing gum made of these sugar alcohols is practically indigestible, and hence, good for our teeth, hopefully. **Aspartame** is a dipeptide of aspartic acid and phenylalanine, which is esterified with methanol at the C-terminus. A peptide chemist could describe it as N-Asp-Phe-Me-C. Because aspartame contains phenylalanine, this sweetener is not suitable for people with phenylketonuria, a hereditary metabolic disease in which the phenylalanine metabolism is not working. Still, aspartame is about 200 times sweeter than sugar and is widely used as a sweetener in food and drinks. Fun fact: aspartame is only one of several very potent sweeteners, all made from peptides. Finally, there is **Stevia**, an extract from the Stevia plant *Stevia rebaudiana*, which is also called sweet herb or honey herb. The main ingredient of its sweet syrup is stevioside, one of several steviol glycosides—all of these substances consist of a three-ring system, which then is glycosylated several times. Although Stevia was known in Latin America for ages, it only recently has started sweetening our lives in Europe—it was the year 2011 when steviol glycosides were approved as a food additive in the European Union.

◘ **Fig. 2.3** **Sweeteners are chemically diverse—here is a small selection. Left**: Sugar alcohols are bulk ingredients in chewing gum. The structure of sorbitol can be seen here. **Middle**: The structure of the dipeptide aspartame—abbreviated to N-Asp-Phe-Me-C—is shown. Enjoy your soft drink. **Right**: One of the main ingredients of the sweet juice of the stevia plant is stevioside. Here I have represented the sugar residues "garnishing" the central building block only schematically. (Image credits: Chewing gum, © ALF photo, ► stock.adobe.com; Soft drink, © Lemonsoup14, ► stock.adobe.com; Stevia, © Daniele Depascale, ► stock.adobe.com)

2

Cheerful Sugars are central stores of energy, but they can't be stored very well—we'll discuss the glycogen solution to this problem later in ▶ Sect. 6.2. There are different ways to classify sugars. Composite sugars tend to be constructed via glycosidic bonds and ester bonds. We highlighted selected sugars—sucrose, glucose, and heparin, as well as dietary fibers. We finished this section with sugar substitutes, and how our guts deal with them. Life has never been sweeter.

Meandering: Sweets and Our Guts

We all have heard it many times: we should eat healthily and exercise regularly... Yes, we should... This protects against diseases and would extend your life, people say. But why is stuff like this so much easier uttered than implemented in real life? Why do so many people have overweight, world-wide? And why do so many dishes, snacks, and sweets taste so delicious that many of us can't get enough of them. Questions to which there are many answers, but certainly no easy ones.

Eating what keeps us healthy (and in shape?), is an ultra-old survival strategy—certainly much older than we humans ourselves. The control circuits for regulating food uptake are deeply ingrained in our brain. Most of this regulation happens subconsciously; it does not need to be learned at all. One of the mechanisms involved is the preference for sweet and the avoidance of bitter; as sweet taste usually stands for calories, but bitter bits normally signal inedible plants and danger.

Modern human societies just face one major challenge in this regard: It is that our diet radically changed at around the middle of the twentieth century. All of a sudden, industrialized nations catapulted themselves into a world of plenty, with food in abundance. What tastes best, does not necessarily be the best for you—we had to learn this slowly. And then artificial sweeteners entered the scene. Research has shown that our tongue cannot distinguish natural sugars and those calorie-free sweeteners. Once swallowed, however, our guts immediately sound the alarm, if calories announced by the tongue turn out to be "empty". Whether we want to or not, our stomach makes us eating sweets. And in the end, there is a single organ in our body with the greatest appetite for sugar, by far, our brain.

For your reading list: The tongue cannot distinguish sweeteners from real sugar, but our gut can—Buchanan et al. **2022**. The preference for sugar over sweetener depends on a gut sensor cell. *Nat Neurosci*. 25(2): 191–200. ▶ https://doi.org/10.1038/s41593-021-00982-7.

Good to know: Sweet receptors are not confined to your tongue, they can be found all over the body, including in the gut and the hypothalamus—Lee & Owyang, Sugars, Sweet Taste Receptors, and Brain Responses. *Review Nutrients*. **2017**; 9(7): 653. ▶ https://doi.org/10.3390/nu9070653.

■ **Navigation**

More on glucose in ► Sect. 2.7.

For the polysaccharide glycogen, there is a dedicated meandering in ► Sect. 6.2.

Aspartame is just one sweet peptide of many! Piero Andrea Temussi is a gourmet when it comes to Sardinian cuisine. He also studies the taste of peptides and proteins—Temussi. **2012**. The good taste of peptides. *J Pept Sci*. 18(2): 73–82. Review. ► https://doi.org/10.1002/psc.1428.

2.2 Amino Acids, Peptides and Proteins

An **amino acid** is a carboxyl **acid** with at least one **amino** group. In organic chemistry, there would be a huge selection of amino acids (please think again about **napkin III**, where we discussed the vastness of the chemical "space"). Biochemically however, proteins are almost exclusively composed of just twenty amino acids. A bit boring, isn't it? Fun fact: Proteins are named after the Greek word *proteus*, for being the "first", or the most "fundamental". Well, in this very ► Chapter 2, they only come second. And later on, we'll learn that they actually come third, after DNA and RNA, all encrypted in the genetic code. This genetic code is called "universal" but still holds surprises. The amino acids selenocysteine and pyrrolysine are such exceptions; clever deviations from the universal code make the ribosome to incorporate these special amino acids; they are described in ► Sect. 3.6.

All amino acids in proteins have an amino group attached to the carbon atom that directly follows the carboxyl group, this C-atom is called α-C-atom or also short C_α. Counting the carboxyl group and an amino group, the α-C-atom has bound two different groups already. Should the other two groups also be differing, this α-C-atom becomes chiral—there is a D- and an L-version of nearly every proteinogenic amino acid. The protein production factory of all cells is the ribosome, and in all known life forms the ribosome exclusively accepts L-amino acids.

In the cell, nothing ever would happen without proteins. So, one may regard **the twenty proteinogenic amino acids** as the alphabet of life. Still, students may have difficulties memorizing the side chains, the abbreviations and possibly also the pK_A values of special side chains of amino acids. Each of the twenty "canonical" amino acids is something special. For each one, there is something to tell about their history of discovery and all sorts of further peculiarities. Would these stories maybe help with your learning? Asparagine, for example, is abundant in asparagus and was first isolated from asparagus juice (◘ Fig. 2.4), and this is how this amino acid got its name from. For fans: After a decent serving of asparagus, there is a typical smell of urine; this however is probably caused by sulfur-containing compounds, not by decomposed asparagine. In this book, only a handful of amino acids are introduced in more detail. My **meandering** around the one-letter code may help with the abbreviations of the protein-forming amino acids.

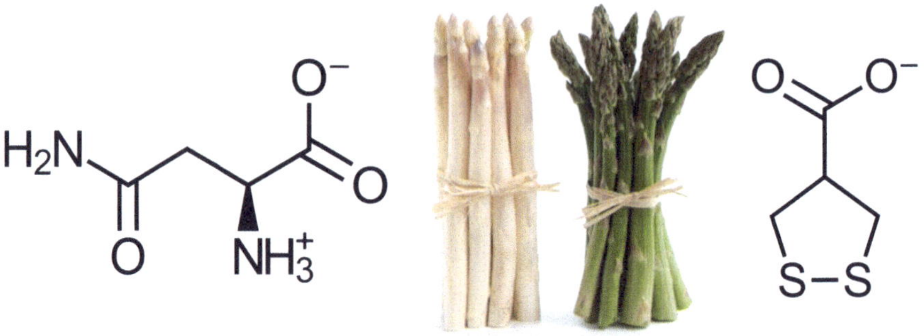

Fig. 2.4 **Asparagine and asparagus have a lot in common**. As the first amino acid ever, asparagine was isolated in pure form. Asparagine is found in abundance in asparagus juice, this amino acid was crystallized therefrom. Asparagine also got its name from asparagus; both depicted in this figure. There is however also a peculiar aroma of our urine associated with the consumption of a good portion of asparagus. This aroma most likely is not caused by degradation products of asparagine; but from the breakdown of sulfur-containing compounds, such as asparagusic acid, which is also shown in the illustration (**right**). An amusing piece of research reported that there are people with and without a "nose" for the special asparagus aroma in urine. The author of this book certainly has that nose and is also a fan of asparagus—Markt et al. **2016**. Sniffing out significant "Pee values": genome wide association study of asparagus anosmia. *Brit Med J*. 355:i6071. ► https://doi.org/10.1136/bmj.i6071. (Image credit: Asparagus, © emuck ► stock.adobe.com)

Meandering: How Can I Memorize the One-Letter Code of Amino Acids?

Why would you probably want to remember which letter of the alphabet stands for which amino acid? We'll discuss this by the end of this text box. Maybe you simply have to... Probably the best way to approach this learning task by working out your own solutions. So, did I. Here is what I did: I divided the amino acids into groups and then developed my own mnemonics. Going through the alphabet, I sorted the letters into three groups. Firstly, the amino acids that have their initial letter as an abbreviation. Secondly, those amino acids that have a different letter as an abbreviation. And finally, those letters that do not stand for an amino acid at all.

The first aha moment could be that the first group almost only contains "simple" aliphatic amino acids. The somewhat "more complicated" amino acids are in the second group—ranging from polar acid amides and aromatic amino acids to properly charged ones. Time to get creative.

While looking at these amino acids and their "more complicated" structures, you can think up mnemonics. In this way, you might succeed in linking the respective one-letter code and the structural formula of these amino acids. Knowledge for a lifetime?!

The abbreviation is the initial letter	Abbreviation is NOT the initial letter	No AA
A—Alanine	D—Aspartate—four C-atoms, fourth letter in the alphabet	B
C—Cysteine	E—Glutamate—five C-atoms, fifth letter in the alphabet	J
G—Glycine	F—Phenylalanine—**FF**enylalanine	O
H—Histidine	K—Lysine—K is the letter before L	U
I—Isoleucine	N—Asparagine—asparagiNNNe is smaller than Qlutamine—N comes before Q	X
L—Leucine	Q—Glutamine—Glutamine is greater than asparagiN—Q comes after N	Z
M—Methionine	R—Arginine—aRRRginine	
P—Proline	W—Tryptophan—tryptoWan, the only amino acid with tWo rings	
S—Serine	Y—Tyrosine—tYYYrosine	
T—Threonine		
V—Valine		

Back to the provocative question from the beginning: Why should one remember the one-letter code of amino acids at all? Answer: It depends on what you want to do. If you are into bioinformatics, it is incredibly informative to know the amino acid alphabet by heart. Protein sequences are stored in various databases in this, and only this format. Maybe even better if you learn to "see" biochemical properties in the individual letters. LFLVLILL looks pretty hydrophobic, doesn't it? On the other hand, EDEERDSD is quite acidic and highly charged. Can you see, what I do?

The twenty standard amino acids are the building material for quite a few **neurotransmitters and hormones**. The amino acid **tyrosine** is the building material for quite a few other secondary natural compounds—some in the human metabolism, others exclusively by plant realm. For the human thyroid hormone to be made, two copies of the amino acid tyrosine are pressed together and then sprinkled with iodine. What could possibly be easier. Making that iodide-loaded tandem tyrosine is anything but easy. We knew for long that that thyroxine, affectionately also called "T4", is made in the thyroid gland. We knew that it is made in some cellular bubble-shaped reaction factories, the follicle.

Fig. 2.5 **Tryptophan and chocolate and serotonin.** Our body uses the amino acid tryptophan to make the neurotransmitter serotonin, among other things. Serotonin brings feelings of well-being and moments of happiness. Could the craving for serotonin then be the reason why so many of us like tryptophan-rich foods… … … such as chocolate. Maybe… More likely, however, just the stuff for a good memory anchor—there is no scientific evidence that additional tryptophan intake in tablet form helps against depression and mood swings. Just by the way, I have depicted tryptophan somewhat twisted, so that the comparison with serotonin is easier. Can you double-check if it still is an L-amino acid in this structural representation? (Photo: Chocolate, © Lumos sp, ▸ stock.adobe.com)

Only recently, however, more molecular details were discovered about T4 biosynthesis; it probably is one of the most complicated biosyntheses known. Within the thyroid gland and in the follicle bubble, T4 is made by and from the thyroglobulin protein. Thyroglobulin is a giant protein, weighing more than 600 kDa. Working in pairs, a dimer of this colossus only features 7 positions, SEVEN, where one thyroxine molecule is formed per site, the rest of the protein is chopped in pieces and degraded thereafter.

Another aromatic amino acid is tryptophan. Something that is made from **tryptophan** can be seen in Fig. 2.5.

Peptide bonds link amino acids together. A peptide bond always extends from the C-terminus of one amino acid to the N-terminus of the next one, making "-CO-NH-" the written representation thereof. A peptide bond is not stubborn, but it is strangely rigid—it simply doesn't rotate as freely as other bonds do (Fig. 2.6). This is unexpected at first glance.

After all, the bond between the carbon of the carbonyl group and the nitrogen of the amino group is represented as a single bond "-CO-<single bond>-NH-". Hence, it should rotate feely. But the peptide bond is not a single bond after all—occasionally some electrons flip back and forth within the "-CO-NH-" system, conferring the peptide bond a bit of a character of a double bond. Think of the peptide bond as a chemical oddity, or not, this bond is present in ALL peptides and proteins. Only the single bonds next to the α-C-atom are freely rotatable; the whole of the peptide bond remains flat, however.

Peptides Must Start and End Somewhere In a longer and linear peptide, an amino group and a carboxyl group inevitably remain free. So, there must be a beginning and an end to that peptide or protein. For a biochemist, **the beginning** clearly is **the N-terminus,** and **the end** is **the C-terminus**. It is exactly this order in which peptides are formed at the ribosome, the cell's large protein factory. Ribosomal proteins always start with a methionine residue at the N-terminus. But be careful when talk-

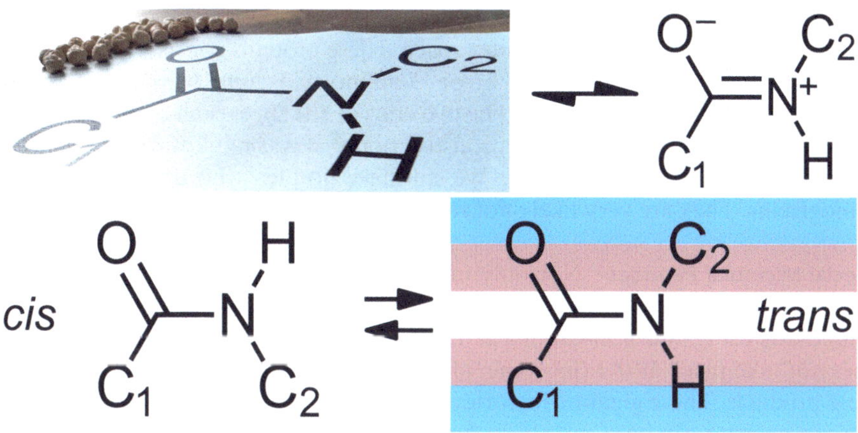

Fig. 2.6 **The peptide bond is flat. It does not freely rotate, it is locked**. The peptide bond is flat as a board. All four atoms of the actual peptide bond as well as the two α-C-atoms (C_1 and C_2) of the involved amino acids lie in one plane. **Top left**, I printed the peptide bond on plain paper and photographed it from a perspective; a few chickpeas are supposed to remind us of protein. Why is the peptide bond flat? First, we push a few electrons back and forth. Nitrogen's free electron pair (not shown) likes to lean over to the carbonyl carbon, just so the electron-hungry oxygen can pull the shared electron pair of the carbonyl group all to itself. **Top right**, this gives a bit of a negative charge to the oxygen O and a bit of a positive one to the nitrogen N. The thing in the middle is not a reaction arrow, it is a double-headed arrow, supposed to suggest resonance—in real life the peptide bond oscillates between two states, it is a bit of a single bond and of a double bond at the same time. To an extent, the peptide bond is a double bond—not completely, but neither not at all. The partial double bond character causes the peptide bond to rotate only very limitedly. And that's why there are two states of the peptide bond. **Bottom line**, thinking of the peptide bond itself as a line halving the structure, if both α-C-atoms are on one side of the bond, then we speak of a *cis*-peptide bond. If they are on opposite sides, then it's a *trans*-bond. In peptides, the bond in *trans* is the rule; *cis* bonds are the exception. (Image source: Peptide bond, JWM, 2022)

ing to synthesis chemists about peptides. Their standard synthesis on small beads or on membranes, the solid phase synthesis, actually starts at the C-terminus. So they might think the other way around.

Oligopeptides, Polypeptides and Proteins Linking two amino acids, a couple of them, or even many amino acids, the resulting polypeptide chain can be short or seemingly endless. Some few linked amino acids are called peptides. **Peptides** are important signaling substances in the human body. The human peptide hormone insulin consists of 51 amino acids, for example. Insulin is a central regulator of blood sugar levels and metabolism overall. Longer peptides are eventually called **proteins**. Personally, I don't think too positively about setting arbitrary boundaries between peptides and proteins. It is probably better to call a longer peptide a protein that can assume an ordered three-dimensional structure. But even this definition has its pitfalls. The structures of "3D-peptides", i.e. proteins, can be vast; there are all sorts of ways in which those "peptide spaghetti" can wind up. Looking at 3D protein structures, you probably can't avoid seeing pretty corkscrews and groups of arrow patterns. We will discuss these α-helices and β-sheets in ► Sect. 4.3.

Exotic Proteins A special class of proteins was recently discovered in samples from meteorites—within matter from outer space. These proteins consist of two peptide chains of glycine residues, each 15, 16 or 17 amino acids long, which form an antiparallel, double-stranded β-sheet. The two ends of the sheet each contain a bridgehead of iron, oxygen and lithium, a construction that was not known before. These proteins were named hemoglycins. We still have no idea who or what assembles hemoglycins. They are very likely not randomly formed.

Special Microbial Peptides Not only in meteorites, but also in bacteria there are extraordinary peptides. Very unusual bacterial natural products are even partly composed, using the non-canonical D-amino acids. Such unusual biosynthesis does not take place at the ribosome; instead, some molecular knitting dolls make these products. These gigantic factories are the non-ribosomal peptide synthetases. Microorganisms can be at relentless chemical war. So, it might represent a strategic advantage to have such synthesis machinery to produce peptide-based antimicrobial peptides. We humans are still in the process of surveying this huge chemical arsenal and to put it in good use for our well-being, as peptidomimetic drugs, or next-generation antibiotics.

Compact Guide All known life uses α-amino acids in their L-form to assemble peptides and proteins at the ribosome. The peptide bond is the special bond that holds peptides together at their core. The peptide bond is not freely rotatable. In the body, amino acids are sometimes used as building material for hormones and signaling substances. We have also discussed the one-letter code of amino acids and have heard about some strange proteins. This section ends with a **meandering** on antibodies in analytics.

Meandering: Antibodies Almost Always Come from the Western Part of the Lab

Antibodies are central proteins from our immune system. They are shaped like the letter Y. At the two short ends "at the top" of the fork-shaped Ypsilon letter are the binding sites for antigens. The biochemical and genetic processes during the maturation of antibodies is complex and would fill several hours of lecture. Here just this much: Clever DNA cutting and somewhat sloppy repairing during DNA recombination, provides a nearly endless sequence diversity in the *binding loops* of the antibodies.

Antibodies can bind almost anything. Well, some antibodies are endowed to bind very unique proteins or other biological structures with great specificity. It is this selectivity or specificity that makes antibodies so interesting for biochemical analytics—for example in the Western Blot. In such a blot, a protein mixture—from a cell or a tissue sample—is separated in a gel (here, we call this gel simply SDS-PAGE), roughly according to molecular size. The proteins are then transferred to a membrane, they are blotted—have you been blotting lately?

The size-sorted proteins now are unfolded and stick to that membrane, openly and nicely accessible. From the outside, a specific antibody can bind. Most commonly, a specific antibody is combined with a more general one that has some features that are easy to detect. This second, more general antibody recognizes the specific antibody; and it brings something with it that can be observed easily, e.g. an enzyme, a radioisotope or a small gold ball. Voila, nothing could be simpler.

The Western Blot is a "standard" method in almost all biochemical laboratories around the world. Nevertheless, there are probably no two working groups that perform their Westerns, down to each and every detail. This is somehow like cooking: potato dumplings, an Indian curry or *sauerkraut*, all are absolute standards in many kitchens; they are however not particularly easy for beginners to get right. Many students suffer from poor Western Blots. Eventually, however, they get better, the Westerns. Definitely. Or so they say. Really.

The Western Blot was indirectly named after Edwin Southern. This is not a joke. Dr. Southern revolutionized molecular biology by detecting specific bits of DNA with labeled DNA snippets or probes. Dr. Southern had invented the Southern Blot. In his honor, the corresponding blot with RNA was then named Northern Blot and then the proteins with the antibodies for the Western came along, where a protein might be detected by a primary antibody (orange), that then is visualized by a secondary antibody (blue), attached to detection technology, be it an enzyme, a gold particle, or some radioactivity (all symbolized by that one orange asterisk*).

There are several deviations of Westerns: Far-Westerns might sandwich at least one more protein in between. South-Westerns might detect some specific DNA-binding protein. Certainly, there are more major Western flavors, and hundreds of subtle variations. Even in eastern labs, there certainly are Western blots. Easter, however, is reserved for the Easter festival and the family.

(Image credit: Compass rose from the Midlands Art Centre in Birmingham, JWM, 2021; SDS-PAGE gel, JWM, 2012)

▪ **Navigation**

More about the genetic code can be found in ▶ Sects. 3.5 and 3.6.

Hemoglycins—proteins from Outer Space. **One** of several manuscripts about the discovery, is this one—McGeoch & McGeoch. **2024**. Sea foam contains hemoglycin from cosmic dust. RSC Adv. 14(50):36919-36929. ▶ https://doi.org/10.1039/d4ra06881e—**and** an early preview article—Mueller JW. **2020**. Protein X: a protein with iron and lithium from a meteorite, Journal Club. *BIOspektrum*. 26(4):411.

For those who like it complicated, here is a preview of the structure of a protein giant, thyroglobulin.—Mueller JW. **2020**. The structure of a protein giant and thyroxine biosynthesis—Thyroglobulin; Journal Club. *BIOspektrum*. 26(2):180.

Thyroglobulin keeps on giving; here is a higher resolution structure from 2022—Adaixo et al. **2022**. Cryo-EM structure of native human thyroglobulin. *Nat Commun*. 13(1):61. ▶ https://doi.org/10.1038/s41467-021-27693-8.

2.3 Fats, Oils and Good Butter

This chapter deals with biological oils and fats, good butter and margarine, and some waxes in the end. Some special lipids are highlighted here; from some of those, we will assemble whole membranes in ▶ Sect. 4.2. To show how much energy is stored in fats, we will burn a peanut, what a lovely and sooty black mess. But first we start with petroleum or mineral oil.

Mineral oil is liquid gold. The industrial world can't get enough of it, wars are fought for it. Mineral oil, also called petroleum, is the result of fossil transformation processes of organic remains of algae, zooplankton, and other stuff. Petroleum consists of linear or branched alkanes, cycloalkanes and aromatics. Alkenes are found relatively rarely, in there. There are almost never alkynes (◘ Fig. 2.7 explains all these terms).

Contrarily, biochemical fats contain considerable amounts of water; mainly hidden in the ester bonds between fatty acids and special building blocks. A conceivable path from biological fats to petroleum is the removal of water from precursor material over geological time scales.

Fats consist of a core component, usually glycerol, and several carboxylic acids with long hydrocarbon chains. The longer carboxylic acids are also called fatty acids. The carboxyl groups of three fatty acids, all are glued to the glycerol hydroxyl groups in **ester bonds**—stable, but still easily cleavable connections. What matters is: all three negative charges of the fatty acids are "disguised", this way these fats received their name—**neutral fats**. Fatty acids in neutral fats mostly have single C-C bonds. A few have C=C double bonds. Literally never you'll find C≡C triple bonds. (Very much like petroleum: many alkanes, rarely alkenes, nearly never alkynes). Biochemical fatty acids rarely have branches in their long fatty alkane or alkene chains. They usually consist of an even number of carbon atoms. All of this can probably be better understood through their biosynthesis, which we discuss in ▶ Sect. 6.5.

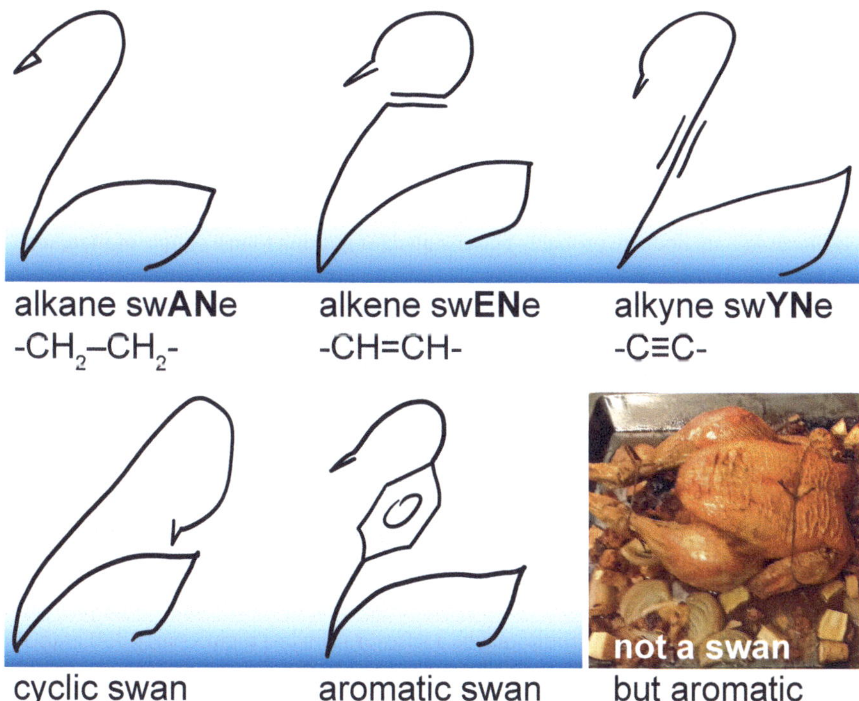

Fig. 2.7 What are alkanes, alkenes, and alkynes, once and for all? An illustration of different hydrocarbons, with the help of swans. Most of these hydrocarbons are long-chain compounds with several C-atoms. There are many variations of this meme. For the author of this book, it all started with a T-shirt showing porcupane, porcupene, and porcupyne. In social media, there was a lot of similar memes around a prominent English football player who then transferred to a South-German football club. You know those memes, some of them ending on harrykaneol. Do you possibly know other, more suited memes/mnemonics, in whatever language? Please share your most creative appropriate contributions with me. (Image credit: Drawing, JWM 2022; Christmas chicken, JWM, 2021)

Despite the one carboxyl group per fatty acid and the occasionally occurring double bonds, the empirical formula of neutral fats with longer fatty acids is still quite close to an alkane with $(CH_2)_n$. The carbon here is at its maximum possible reduction stage. One can already guess that fatty acids store a lot of energy—there is a lot to oxidize or to burn here. But how much energy is actually in, say, a peanut?

This is an experiment that can be seen on the internet in various forms. A wonderfully sooty mess. Comparable experiments are also available with all sorts of other natural substances or foods. How this works is actually really easy. On **Napkin IV** you will get a rough idea of how much energy such an experiment might reveal.

2

Napkin IV: How Much Energy is in a Peanut

The experimental setup vaguely represents a calorimeter. A peanut is fixed somehow, perhaps with a pair of pliers, a wire rack or a bent paper clip. Above it, a test tube is attached, containing any quantity of water that you measured precisely, 10 g of water in our example. Measure the temperature and the amount of water before and after (weigh!) and write it down. Bunsen burners or blowtorches (yes, those for a *crème brûlée* are fine) are probably better suited for igniting than a couple of matches.

What will happen? Let's take a closer look at the table of nutritional values printed on the back of a pack of peanuts, maybe a leftover from last night's party. 100 grams of roasted, unsalted peanuts may be listed to contain 23.7 g protein, 21.5 g carbohydrates, 8.0 g fiber and almost half of its weight in fat, 49.7 g maybe. For a total of 100 g of peanuts, the caloric value is about 585 kcal or 2449 kJ.

The single peanut in our experiment weighs about 1.4 g, so we have a maximum of 8.19 kcal of energy available. According to the definition of the kilo calorie, this energy could heat up to 8.19 kg of water by one degree Celsius. In other words, a single kcal of it would be enough to heat our 10 g of water from room temperature to boiling.

That leaves about 7 kcal to go. Making steam out of water is extremely energy-demanding at 539 kcal/kg or 2257 kJ/kg. Still, a good 5 kcal would be needed to completely evaporate our 10 g of water.

Of course, burning that peanut as well as the transfer of heat from the burning nut to the test tube is not optimal. Most likely, the result will be that the liquid has been heated by perhaps 60 degrees Celsius and that 1–2 ml of water have evaporated. Nevertheless, it is quite impressive what is contained in such a small peanut.

What is described here is more or less the principle in which calorimeters function. In those machines all aspects of the process are optimized, so that food scientists can determine those energies much more precisely.

(Photo credits: Peanuts © photocrew, ► stock.adobe.com; fire © JWM, 2022)

About Fats and Oils What is regarded a fat and what is an oil? The answer mainly depends on the melting temperature of the respective fat/oil and the ambient temperature. Think of a good olive oil that goes strangely solid, maybe in the fridge or even in a cold car, when driving home from grocery shopping. Some oil becoming a solid, is a completely harmless process—nothing is wrong with your precious olive oil. In a warm living room, the oil will get liquid again just as quickly. Otherwise, the same relationship applies as with alkanes: The longer the carbon chain, the higher the melting temperature. The number of double bonds also affects the melting temperature of fats. Double bonds create a kink in the otherwise quite straight tail of fatty acids and thus prevent a denser packing. Many double bonds lower the melting temperature—we are back with an oil again.

And then, there is endless discussion about whether **"good" butter** or **margarine** would be better for you. Butter is a natural product with a wide range of different fatty acids, and a content of unsaturated fatty acids similar to other dietary products. Butter naturally contains a high proportion of the fat-soluble vitamins A, D, and E. Margarine is an industrial product that used to be quite nasty in the past. The chemical hardening of vegetable oils led to a high proportion of unhealthy trans fats. Margarines have certainly improved nowadays—I can't believe it's not butter. They are also fortified with vitamins. Which of the two would be the better choice, is really hard to say. Both still are high in calories. Both should probably be used only sparingly. Why not also use other dietary products, tomato puree, or hummus as a spread on your bread?

A special class of fats are **waxes**. These are quite long-chain fatty acids, linked by ester bonds to alcohols that are distinctively different to glycerol. This time, these alcohols come with only one OH group; waxes also grow very long in their hydrocarbon chain. Waxes are perfect for waterproofing feathers and other surfaces. Several bird species have centralized their wax production in the preen gland. Waxes also serve as building material for some insects. A major ingredient in bee-made wax is palmitic acid-myricyl ester CH_3-$(CH_2)_{14}$-CO-O-$(CH_2)_{29}$-CH_3, which consists of the C_{16} fatty acid palmitic acid and that "never-ending", long C_{30} alcohol myricin—this wax is a hydrophobic giant with 46 C-atoms!

Other Types of Lipids If you want, please use "**lipids**" as the collective term for everything fatty in the human body. "Fats" then only mean a sub-group group of lipids, namely the **neutral fats**. Other types of lipids are made from neutral fats; they have one of the fatty acids swapped for a phosphate group—the **glycerophospholipids**. Just like a free fatty acid, these lipids are amphipathic—they have a water-loving and a fat-loving part. Amphipathic lipids can form micelles or even merge to form membranes. Please think of soap bubbles now—in a way, they are membranes, but the other way around (► Fig. 4.3). **Sphingolipids** certainly are something special. They don't contain glycerol as a scaffold, instead they incorporate the slightly longer alcohol sphingosine. And there are **sterols**, these are lipids related to steroids. Cholesterol is probably the most famous sterol; it regulates the behavior of membranes in many ways.

Clarification Fats are a large class of natural substances. Fats are indispensable for energy storage. They also serve as building blocks, buffering substances, and insulating materials. Fats are made out of some sort of alcohol, mostly glycerol or sphingosine, and fatty acids. Fatty acids are long, unbranched α-carboxylic acids. Lipids not only include fats; they also include amphipathic molecules that act as soaps; these soapy molecules can form vesicles and membranes.

2

▪ **Navigation**

You will read more about biological membranes and the role of cholesterol in ▶ Sect. 2.5 on isoprenoids and then again in ▶ Sect. 4.2 on biological membranes.

In ▶ Sect. 5.4 I will discuss fat-soluble vitamins and cofactors; I will also mention Vitamin D.

We will better understand some peculiarities of the long, unbranched α-carboxylic acids when we look at the biosynthesis of these fatty acids in ▶ Sect. 6.5.

However, entire classes of lipids are not mentioned here, for example the **inositols**, involved in various signal transmission processes, the important **sulfatides**, mainly found in the brain, or even **sulfated neurosteroids**.

2.4 Nucleic Acids

In other contexts, you certainly have heard of carbohydrates ("sugars"), proteins, and fats as types of natural substances already. Another major class of natural substances—the **nucleic acids**—might be less well known, even in spite their many and diverse biological functions. We will look into the nucleic acid content of some types of food. The big biochemical difference between nucleosides and nucleotides will be highlighted. And we will delve into some early episodes of the discovery history of DNA, which happens to happen in Germany.

"Nucleic acids" might only be familiar to some, but everyone has heard of DNA. The abbreviation DNA might be as well known in the general population as the formula H_2O for water, ATP as cellular energy currency, or Einstein's formula $E = m \cdot c^2$. In the corporate world, the phrase "it's in our DNA" has become synonymous with core values, good character, or so... So be good for DNA's sake.

As a class of natural substances, nevertheless, nucleic acids are less in the general perception. Perhaps, this is because they are composite natural substances. But that's a weak argument, as neutral fats are composites as well. A contributing factor might also be that nucleic acids do not appear in today's nutritional value information boxes on food packages—those "traffic lights" only feature other information. Still, nucleic acids are an important part of our diet. They are in every food. And yes, nucleic acids also contain calories. **Napkin V** shows how many nucleic acids we eat every day.

Napkin V: How Much DNA Do We Actually Eat Per Day?

Each and every naturally grown food contains nucleic acids, trust me. If you see a "DNA-free" or "gene-free" sticker on food somewhere, keep your hands off that stuff. Instead you could go into a discussion about next-generation breeding techniques, including CRISPR—we will briefly discuss these topics in ▶ Chapter 7.

A whole fish could have less than **1 g DNA per kg** dry weight. [I will use "DNA" and "nucleic acids" somewhat interchangeably here. This is just an approximation. There is always somewhat more "nucleic" acids than DNA, when also quantifying RNAs, free nucleotides, and nucleotide cofactors].

Muscle meat has significantly more DNA, maybe **7–9 g/kg** dry weight. Offal can contain quite a lot of DNA. Liver can reach up to **20 g/kg** dry weight. Baker's yeast (**6 g/kg**) and different varieties of cabbage and have a very similar content of DNA. The ever so healthy broccoli (**5 g/kg**) comes, for vegetables, with a fairly high content of nucleic acids.

I didn't live in the urban Ruhr area, the *Ruhrpott*, for 11 years without learning to appreciate a proper *currywurst*. A serving of this German kitchen classic may contain 150 g of currywurst, 150 g of fries and as much delicious sauce as you like. Let's translate this into more or less finely minced pork, best potatoes and sauce, according to family recipe, i.e., tomato, onion, sugar and all sorts of delicious spices.

For our DNA-dietary considerations, let's ignore the sauce for now and focus on meat and fries. 7 g of DNA per kg of dry weight for muscle meat, it is then. With a higher content of pork belly, it even could be a little less. Optimistically, the water content of the sausage is estimated at 20%. That's **0.84 grams** of DNA in the sausage.

For the fries, we are 1 g of DNA per kg of dry weight for fresh potatoes. A water content of 80%, brings us to **0.03 grams** of DNA in the fries.

All added up, we only get **0.87 grams of DNA** for a serving of currywurst.

It becomes clear that stating the calories for the little amounts of nucleic acids contained, makes little sense for food labeling. How about gout then? That tiny bit of DNA compared to the other nutrients can't really influence if one develops gout, can it? Well, it probably comes down to quantities. Even if overall nucleic acid content is low, animal-based food still seems to contain quite a lot of nucleic acids, compared to plant-based food. And since our body only slowly breaks down the purine bases adenine and guanine, a little more salad, fruit or vegetables might not really harm those suffering from gout. Maybe a diet that is good for all of us.

More information can be found here: A wonderful table on the content of nucleic acids in different foods.—Jonas et al. **2001.** Safety considerations of DNA in food. *Ann Nutr Metab*. 45(6): 235–54. Review. ► https://doi.org/10.1159/000046734.

People on a vegetarian diet develop gout less often—Chiu et al. **2019**. Vegetarian diet and risk of gout in two separate prospective cohort studies. *Clin Nutr*. 39(3): 837–844. ► https://doi.org/10.1016/j.clnu.2019.03.016.

Nucleic acids generally come in two "flavors", the nucleosides and the nucleotides. A nucleoside consists of a nucleobase or something similar, and a sugar with 5 carbon atoms, a C5 sugar, or something similar. In addition, nucleotides contain any number of phosphate residues.

Let me try to explain somehow by meandering a bit: Do you know the terms stalactite and stalagmite? Or have you ever wondered why an ordinary mortal should remember different words for hanging and standing drip-stones? Well, with various supporting mnemonics, it actually is not that difficult to keep these two terms apart. How about this one: "StalacTites hang from the ceiling and StalagMites stand on the mat".

ℹ If you have useful mnemonic for distinguishing between nucleosides and nucleotides, please contact me.

Compact Why the categories nucleosides and nucleotides make biological sense, is examined in the corresponding **meandering**. A nucleoside is a sugar or something similar, linked with a base or something similar. There are many such combinations, far more than those building blocks that are built into RNA and DNA. The chemist can think here of the entire chemical space. Significant nucleosides are highly effective drugs against cancer or viral infections. Many of us have probably used acyclovir against cold sores at some point, a really magic remedy against herpes, isn't it? Only nucleotides, not nucleosides, are **components of the nucleic acids DNA and RNA**.

Meandering: When a Nucleoside is Knighted to a Nucleotide

Chemically, a **nucleoside** is rather ill defined. Only phosphorylation turns this "thing"—the nucleoside of whatever kind—into a **nucleotide**. The important biochemical difference is that only something phosphorylated, some sort of a nucleotide, could eventually be incorporated into DNA. And this initial phosphorylation of a nucleoside is often slow and laborious in the cell—it is not that easy to get this accolade. The **Nucleotide** Knighthood does not come effortless. Once a single phosphate is transferred, however, more phosphates can be pushed back and forth almost friction-less between different nucleotides. Somewhat similar to real life, once you have made it, further honors and accolades almost inevitably come.

How well a nucleoside drug is upgraded to nucleotide status, thus determines the effectiveness of this drug. This fact has led to seemingly desperate attempts in drug discovery, such as trying to smuggle-in, to "deliver", a whole gene for a nucleoside kinase into the target cells, alongside the nucleoside drug. The **illustration** in this meandering shows the antiviral drug acyclovir and the normal nucleoside adenosine. For a nucleoside to become mono-phosphorylated, in a way to get their first accolade, can be a tedious process. The accompanying Playmobil scene is to remind us that getting a knighthood sometimes is not easy neither.

(Image information: Knighthood, JWM, 2022)

For a long time, it was considered impossible to simply smuggle some sort of nucleoTide into the cell; we are not talking of a nucleoSide anymore, or some other sort of precursor. After all, the multiply negatively charged nucleoTide molecules can't be membrane-permeable, can they? A new clever way of packaging the nucleotides' precious first phosphate has revolutionized the whole field of nucleoside analogs. The packaging method is called ProTide and allows the direct delivery of a nucleoTide analog—a knight in disguise—into the cell. ProTide is an acronym made of PROdrug & nucleoTIDE. For many nucleoside analogs, the ProTide technology has led to huge increases in effectiveness. This delivery method made a new class of substances to evolve around acyclovir and similar drugs, all significantly more effective against viruses or tumors. There is hope. And, off the record, who needs knights anymore, these days.

▪ For further reading

Manfred Konrad tries to deliver both a nucleoside drug and a nucleoside kinase into the cell, at the same time—McSorley et al. 2014. A designed equine herpes thymidine kinase (EHV4 TK) variant improves ganciclovir-induced cell-killing. Biochem Ph*armacol.* 87(3): 435–44. ▶ https://doi.org/10.1016/j.bcp.2013.11.011.

Youcef Mehellou is very much in tune with the ProTide-Prodrug technology—Mehellou et al. **2018**. The ProTide Prodrug Technology: From the Concept to the Clinic. J Med Chem. 61(6): 2211–2226. Review. ▶ https://doi.org/10.1021/acs.jmedchem.7b00734.

Nucleotides Have Diverse Functions Nucleotides serve as the **energy currency** of the cell—ATP primarily, but honorable mentions also go to other nucleotides like GTP and UTP. Nucleotides mediate between anabolism and catabolism. Chemically, the energy of ATP is hidden in its anhydride bonds between the individual phosphounits. Nucleotides are **part of various enzyme cofactors** (NAD, FAD) and activated precursors (SAM, CoA, PAPS, UDP-Glucose, CDP-diacylglycerol). You can referred to the nucleotide part of these as a molecular handle—a convenient grip where different *business ends* can be attached to.

Nucleotides are important **intracellular messengers**, so-called *second messengers*. Cells might produce these intracellular signals, when they are tickled—stimulated, by hormones or other extracellular signals. One of these *second messengers* is cyclic adenosine monophosphate cAMP. It is produced from ATP by adenylate cyclase and can quickly be broken down by phosphodiesterase enzymes—ideal conditions for an effective signaling system with chemical on/off switches.

Finally, nucleotides are components of the long and linear nucleic acids **DNA and RNA**. These are dealt with in detail in ▶ Chapter 3. The discovery history of DNA is exciting and has partly been written down in book format. A short version of the first part can be read in the following **historical meandering**.

2

Meandering: The Race for the DNA Structure—Part 1

In 1869, this story starts with the primordial ooze, well, more exactly with the disgusting slimy gore that Friedrich Miescher squeezed out from wetting wound dressings, all in the kitchen of Tübingen castle, Germany. The procedure is somewhat related to what almost every visitor to a molecular biology outreach event does these days—isolating DNA. These days however, we don't use oozing and dripping bandages, anymore, but some friendly fruit such as banana, strawberry or kiwi—thanks to modern science communication. Miescher named the old slime "nuclein" and described it as the acidic part of the cell nucleus. He also demonstrated that phosphorus was contained in nuclein. Miescher had discovered DNA. THE discoverer of the genetic substance described his experimental success at this time as follows:

> Chance did not exactly favor me during the time of my investigation. (Friedrich Miescher)

Not many who have completed an experimental thesis can probably say the opposite about themselves. What a misjudgment by Doctor Miescher! The discovery of DNA—a German success story, so far.

The function of the nuclein, however, remained obscure for Miescher—he thought that proteins were responsible for inheritance. And Miescher was not alone with this opinion. When it comes to inheritance, even in the 1930s, proteins were trusted more with their complex composition rather than the "boring", homogeneously composed nuclein. But then various discoveries were made, providing more and more information about the actual nature of the nuclein, which had been renamed to DNA, by then.

For further reading

Friedrich Miescher tells us a lot about the discovery of DNA, quite prosaically, but also somewhat long-winded, at times—Miescher. **1871**. About the chemical composition of pus cells. *Hoppe-Seyler's medical-chemical investigations*. Issue 4: 441–460.

Navigation

▶ Chapter 3 deals with DNA and RNA in greater detail.

The idea of the nucleotide as a molecular handle will be extended in ▶ Sect. 5.3 about cofactors and the special chemistry they allow.

More about ATP and high-energy compounds later in ▶ Sect. 5.6.

2.5 Isoprenoids, Steroids, and Intricate Secondary Natural Substances

Isoprenoids are the last category of natural substances that we will discuss here. Only two different assembly units, each of five carbon atoms—C_5—are combined (we'll just call them IPP and DMAPP) into near-arbitrarily complicated molecules. Admittedly, bacteria and animals can produce a few isoprenoids. However, the true

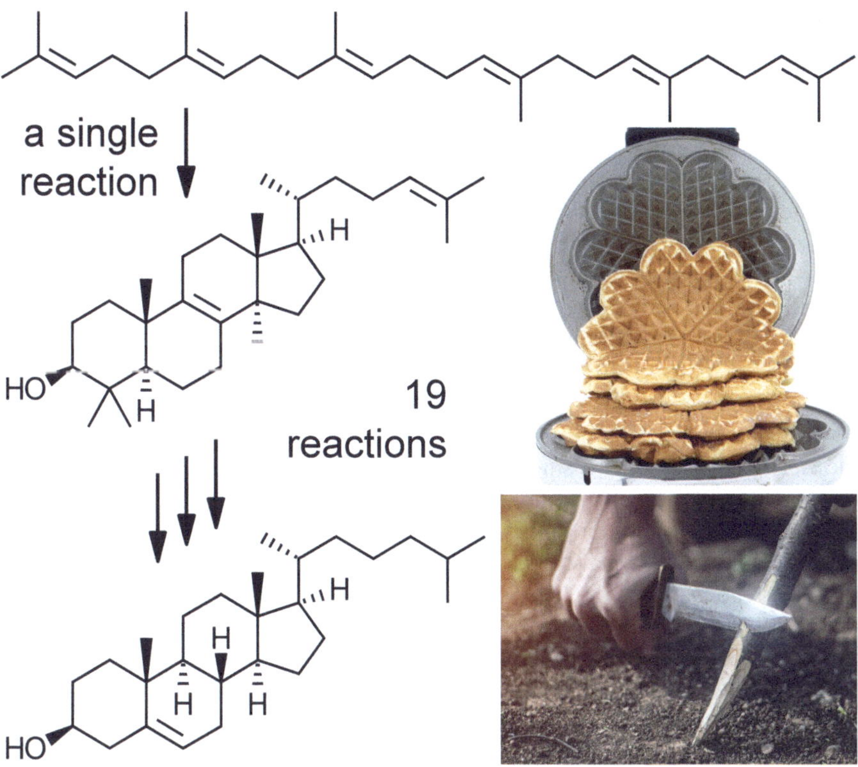

Fig. 2.8 Squalene is "baked" into lanosterol, and cholesterol is "whittled" down from lanosterol. Top: Squalene is an isoprenoid. In a single reaction, it is precisely converted to lanosterol—this exergonic reaction establishes seven chiral centers (**middle left**). 19 more reactions are necessary to produce cholesterol (**bottom left**). One needs to look closely to spot the differences in the two formulas for lanosterol and cholesterol? A traditional German waffle iron represents the action of lanosterol synthase. All the small reactions vaguely remind the author of whittling wood with a carving knife. ► Fig. 5.8 showcases some steroid products, all made out of cholesterol. (Image credit: Waffles, © TwilightArtPictures ► stock.adobe.com; Wood carving, © Marko Duca ► stock.adobe.com)

masters at producing elaborate isoprene structures are plants, as defense substances or fragrance. The structure of the C_{10} molecule menthol, for example, doesn't look like a lot at first glance. However, menthol is very refreshing, when you smell peppermint, make an infusion from it, or just enjoy some menthol drops. Delicious.

The master class of isoprenoids probably are the steroids. Two molecules, each made of three C_5 units, are combined to a carbon chain with 30 C-atoms, the squalene molecule (Fig. 2.8). Squalene is then activated by oxidation to an epoxy, and then something amazing happens: Carefully watched by, well, catalyzed by the enzyme lanosterol synthase (also called oxido-squalene cyclase), lanosterol forms, a molecule with four rings and seven chiral centers. The reaction is strongly exergonic and proceeds more or less on its own. Lanosterol synthase merely keeps an eye on the stereochemistry—pretty much like a waffle iron, pressing the dough into shape. You can already appreciate the basic structure of a steroid—three six-membered rings and one five-membered ring. With almost five rings, one could nominate the baking of lanosterol "waffles" as a new discipline at the next Olympic Games.

The steroid precursor made of carbon composite material is ready. Now it "just" takes some more 19 reaction steps until we arrive at **cholesterol**, the first isoprene masterpiece. On the way, a methyl group is removed here, a double bond is shifted there. Cholesterol is extremely important—as a storage lipid and as a component of membranes. Cholesterol can simultaneously stabilize membranes and keep them flexible. Cholesterol is also the starting material to produce steroid hormones. Biochemically, cholesterol is important enough to us, that "we humans" have still retained the ability to produce cholesterol ourselves—on top of all the cholesterol we constantly absorb from our food.

Almost all carbon atoms in lanosterol and cholesterol are sp3-hybridized, meaning that four single bonds extend from each carbon in a tetrahedral shape. This very arrangement is also found in a diamond. Hence, why not describe the basic steroid structure as a single-layer (of a) diamond. The stability of the diamond and its energetic minimum may serve as an explanation why there are steroids or steroid-like compounds in so many organisms. Could all these steroid scaffolds possibly even be an example of convergent chemical evolution? I leave this to the reader to decide. Here, I merely provide a brief overview of individual representatives of the various substance classes (◘ Fig. 2.9).

A lot of bacteria cannot produce steroids themselves. Many of them can metabolize steroids, though, turning our intestines into an indirect steroid factory of some kind. Interestingly, Mycoplasma bacteria can produce steroids themselves—but there is a suspicion that they stole the necessary genes for this pathway from us mammals. Oh, and then there are bacterial **hopanoids**. They look almost like steroids, but have five rings instead of four—hopanoids truly must be Olympic.

Then we have fungi. They synthesize **ergosterol** and use it—similarly to our cholesterol—to make their membranes more stable.

Insects are funny. They rarely are able to synthesize cholesterol on their own. Still, they are strongly dependent on cholesterol, as they produce their precious molting hormone **ecdysone** out of it. So, insects have to take in cholesterol with their food—or at least some form of sterol, which they can convert into cholesterol then. In this context, I want to mention plant-eating caterpillars and small bloodsuckers, buzzing around us, outside, on a summerly evening.

And then there are plants themselves. Biochemically speaking, they can usually do whatever they "want". Plants take the basic steroid structure—the one with the three six-membered rings and the five-membered ring, at times—and make something new out of it. They oxidize the middle six-membered ring, affectionately called the "B-ring". And they oxidize that B-ring again and suddenly it has become a seven-membered ring. We have arrived at the **brassinosteroids**, which are a large class of plant isoprene substances. Most of them have this strange internal ester bond in the B-ring, which also can be called a lactone.

Compressed Almost all carbon atoms in all the sterol or sterol-like compounds discussed here are sp3 hybridized, meaning that four single bonds extend from them in a tetrahedral shape. This arrangement is also found in diamonds. The basic

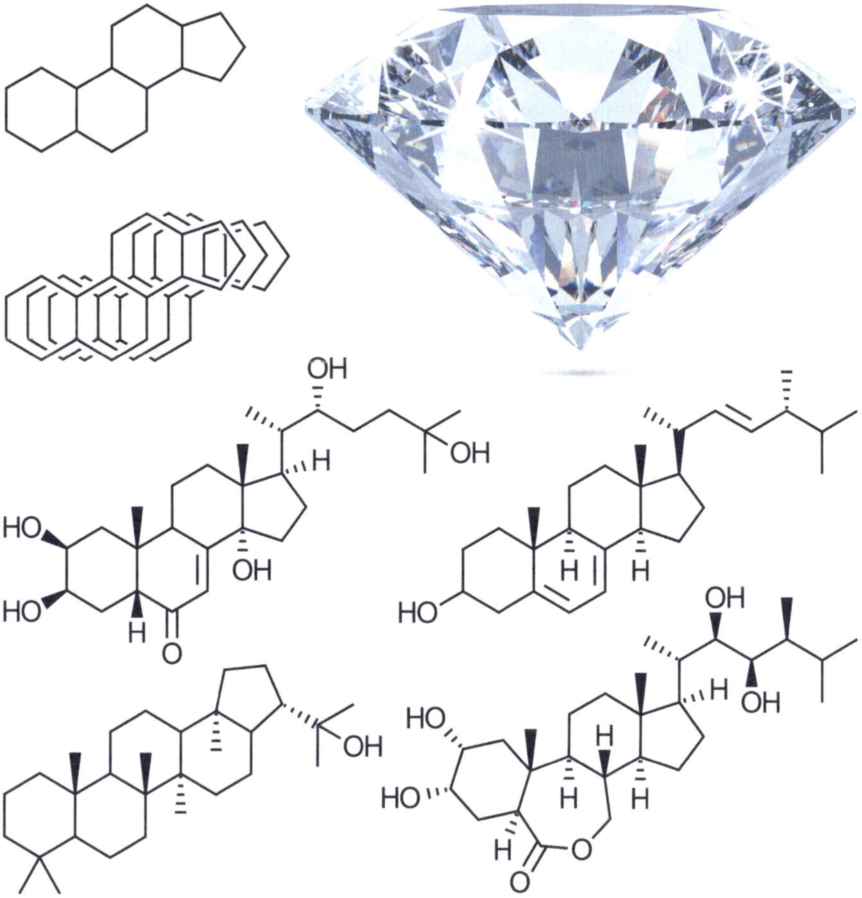

■ **Fig. 2.9** **Steroids are everywhere**. **Top left**: The steroid basic structure is shown, once alone, once stacked three times. This building block can be aptly described as a single-layer diamond. Steroids are everywhere. Pictured are: **Middle left**, ecdysone from insects. **Middle right**, Ergosterol from fungi. **Bottom left**, diplopterol, a bacterial hopanoid. **Bottom right**, brassinolide, a plant brassinosteroid. More on this in the main text. We have already discussed animal cholesterol. Human steroid hormones and vitamin D3 are covered in ► Sect. 5.4. (Image credit: Diamond, © www3d ► stock.adobe.com)

steroid structure can therefore be compared to a single-layer diamond. When we think about the stability of diamonds, we quickly arrive at energetic minima to explain why we find very similar steroid structures in nature, again and again. More on this in the following chapter.

■ Navigation

Cholesterol is a sterol and is similar to fungal ergosterol. Cholesterol is hard-wired into supporting our membranes. See ► Sect. 4.2.

Our body can make biologically highly active steroid hormones from cholesterol. We will get to those in ▶ Sect. 5.5.

Two impressive studies on the presence of various sterol-processing enzymes in various species—Desmond & Gribaldo. **2009**. Phylogenomics of sterol synthesis: insights into the origin, evolution, and diversity of a key eukaryotic feature. *Genome Biology and Evolution*. 1: 364–81. ▶ https://doi.org/10.1093/gbe/evp036—**and**—Markov et al. **2009**. Independent elaboration of steroid hormone signaling pathways in metazoans. *PNAS*. 106(29):11913-8. ▶ https://doi.org/10.1073/pnas.0812138106.

2.6 Looking Backwards Deep into the Times of Prebiotic Chemistry

Nature is diverse. So far, we have discussed sugars, proteins, fats, nucleic acids, and isoprenoids, in this chapter. For all these categories, we have covered general properties and highlighted a few typical representatives. Some natural substances, however, are not yet covered by these five categories. A prominent molecule that falls through this grid is alcohol, or ethanol. Is it some kind of sugar because it is a sugar degradation product? Or is it some weird form of a fatty acid because of its oxidation status? Or perhaps an isoprenoid, as the CH_3-CH_2 bit in ethanol vaguely reminds us of one of the isoprenoid precursors? For now, you would have to consult organic chemistry textbooks to settle on this debate.

Here we want to consider two things. Firstly: Why do we (this time, "we" stands for all living organisms on the planet) use exactly the building blocks in biochemistry that we use. And secondly: Why do we often use exactly those building blocks only in one version—i.e., only as left-handed, or only as right-handed version. Some sort of a spoiler: This type of questions is not easy to answer.

We will discuss energetic minima, spontaneously arising compounds, and frozen accidents. All these events have shaped life, well, at a time before there was any life that we could look back on, through fossils or DNA and protein sequences. It is the **prebiotic chemistry**—the chemistry before biology—that allows us to investigate the origin of life a little closer. Prebiotic chemistry tries to mimic the ingredient cocktail and the reaction conditions that might have been around 3.5 billion years ago. These mixtures are then cooked, sometimes short and hot, sometimes not so hot, but longer. Whatever might have happened spontaneously in the pot, is carefully analyzed finally. Every now and then, this branch of synthetic chemistry provides astonishing answers.

Let's start with looking at the **glucose** molecule in 3D. Glucose has all OH groups well stretched out in equatorial position, no OH group stands bulky axially up or down. Relative to all other reducing six-carbon sugars (some also call them hexo-aldoses), glucose is the single most stable molecule. Overall, one would simply have to recite all possible reducing hexoses, look at them in one or the other projection, and one could conclude that glucose is a good choice, perhaps even the best (◘ Fig. 2.10).

Fig. 2.10 Glucose—Part Two—A local energetic minimum. A ball-and-stick model of beta-D-glucose. On the right end, you can see the OH group that could form a glycosidic bond—the OH group of this hemi-acetal (for fans). Importantly, in this model ALL OH groups and also the six-membered CH_2-OH unit are equatorial—they all point comfortably outward, all almost in a single plane. Not a single group is in an axial position; nothing points up or down, towards the poles. If you look at all conceivable C6 sugars, you might come to the conclusion that glucose is a good choice, an energetic minimum from a limited field of candidates. (Figure index: Model of glucose sugar molecule, JWM, 2022; Globe, © Composer ► stock.adobe.com)

Even the use of α- and β-glucose can be explained reasonably well. The slightly less stable α-glucose is present in the short-lived polysaccharides starch and glycogen. The more stable β-glucose, on the other hand, is found in cellulose and other fibers, durable almost for eternity. Out of a limited field of candidates, glucose is simply the best, a local energetic minimum.

Then there are the **compounds that simply arise** when you stir a few ingredients together and wait for a while, or heat it all up, or both. It may help if some lightning strikes. Actually, we are discussing again most stable alternatives here; this time, however, the observed compounds come out of a huge pot of chemical possibilities. A modern example is this: melt some granulated sugar in a pan or at least heat some fat and then heat tomato paste or diced soup vegetables in it—it sizzles, and wonderful flavors and roasted aromas are created, they are connections of sugars and amino acids. If you mix these ingredients in this combination, similar aromas are created, each and every time. Should you have learned to like the taste, you can learn more about this special kitchen chemistry, by reading around the keyword Maillard reaction.

For many other biochemical building blocks, none of these explanations might apply. How about **frozen accidents**, then? Whatever LUCA, the last universal common ancestor of all our life, biochemically had installed—be it top notch or a "temporary" fix—all descendants have been using exactly these building blocks ever since. A comparison to illustrate this point: In 1949, the company LEGO began producing its famous building blocks. ALL blocks ever made fit all other LEGO blocks, even with perfect fitting options to a block series that is exactly twice as large, Duplo. It is unthinkable to produce these blocks, suddenly differently—say 10 percent larger, or so! And back to our topic from above, can we really causally understand why the standard 4x2 block has exactly the dimensions it has? For now, we must assume "frozen accident" when one variant has prevailed over completely equivalent alternatives, apparently simply by chance.

2

The **primordial soup experiment** by Miller and Urey (1953) was the first spectacular experiment of prebiotic chemistry. A lot of mini lightning bolts were fired, between two electrodes, within a mixture of water, methane, ammonia, and hydrogen. Analyzing the resulting potpourri, the authors clearly detected the amino acids glycine, α-alanine, and β-alanine. They were not quite sure about aspartate and α-amino-butyric acid, though. The message of the experiment was clear: Quite a few amino acids, possibly also other biochemical building blocks, form spontaneously and abiotically (without living beings) from simpler components.

Since then, researchers have been discussing whether this experiment really mimicked the early Earth, or whether there could be more realistic conditions. Also, chemists are trying to produce more and more biochemical building blocks under reaction conditions, that still might be deemed plausibly primordial. Hydrogen cyanide seems to have been of use for making complex nitrogen-containing heterocycles. It's somewhat ironic that something that we only know as a deadly poison today, could have played an important role in the origin of life. For fans: Hydrogen cyanide is called *blausäure* in German—the Blue Acid—isn't this nice?

Classic Prebiotic research keeps on giving. For example, we now know a possible abiotic synthetic pathway to phosphorylated pyrimidine nucleotides under conditions that probably prevailed on the early Earth. A major challenge in this field now is to integrate many different studies with sometimes quite specific conditions into a single framework, a single primordial soup, and then to trace the emergence of more complex biopolymers and even simple cells.

▪ **Navigation**

Mr. Sutherland summarizes a lot of research and concludes that life emerged out of the blue, that is, from cyanide—Sutherland. **2016**. The Origin of Life—Out of the Blue. *Angew Chem Int Ed Engl*. 55(1): 104–21. Review. ▶ https://doi.org/10.1002/anie.201506585.

A research team from England, Poland, and the Czech Republic has succeeded—prebiotically plausibly—not only to produce pyrimidine nucleosides, but even pyrimidine nucleotides—Xu et al. **2017**. A prebiotically plausible synthesis of pyrimidine β-ribonucleosides and their phosphate derivatives involving photoanomerization. *Nat Chem*. 9(4): 303–9. ▶ https://doi.org/10.1038/nchem.2664.

Finally, a few thoughts on how the very first biopolymers could have originated—Adamala et al. **2014**. Open questions in origin of life: experimental studies on the origin of nucleic acids and proteins with specific and functional sequences by a chemical synthetic biology approach. *Comput Struct Biotechnol J*. 9:e201402004. ▶ https://doi.org/10.5936/csbj.201402004.

...and how we could increasingly build cells ourselves synthetically—Gaut & Adamala. **2021**. Reconstituting Natural Cell Elements in Synthetic Cells. *Adv Biol (Weinh)*. 5(3): e2000188. Review. ▶ https://doi.org/10.1002/adbi.202000188.

2.7 How Chirality in Biochemistry Could Have Originated

Understanding prebiotic chemistry becomes significantly more difficult, when you start asking why we use one mirror image of a molecule, but not the other one. Maybe you have noticed, that in ▶ Sect. 2.6 there were no capitalized "D"s and "L"s in any of the substances, unlike to the chapters before. Molecules that are mirror images to another are called enantiomers. Reasonable explanations might include tiny enantio-excesses in otherwise completely 1:1 L:D-mixed mixtures (so-called racemates) that arise from **crystallization**: Imagine: We look at a small puddle, somewhere at the volcanic seashore at low tide. The sun is shining. And suddenly the L-amino acid crystallizes a tiny little better than their D-counterparts. That one is going to be washed away, during the next high tide. …this then happens again and again. Considering a certain time, and we did have quite a bit of time on early Earth, even pure enantiomers could have accumulated this way.

In this book, I try to limit the use of technical jargon to a bare minimum. But here I cannot resist: At least conceptually, we want to try to explain how **homochiral biopolymers** might have originated. Homochiral biopolymers, doesn't that sound like wonderful technical jargon? Almost as if you write it like this: 同手性生物聚合物. This is supposed to be Chinese now, the meaning "homochiral biopolymer" should stay the same—it now is however cryptic to some of us. Starting from a mixture of basic building blocks, longer chains (polymers) are assembled. Once one starts doing something with these chains, for example folding a protein out of it, it becomes clear that a random mixture of D- and L-building blocks is not a good idea. Even a single D-amino acid would disrupt the folding of an entire protein, maybe to a similar extent as a single reversed chain link would do in your necklace—it simply doesn't fit. Now we have two possible roots for homochirality already—a possible enantiomer excess in the primordial ooze in which the basic building blocks were produced, and a functional reason for when longer biopolymers are assembled.

A wonderful story about **snails** provides a third reason for homochirality. Already as a schoolboy, I learned about right-handed and left-handed snails, back in the days. This was about the vineyard snail *Helix pomatia*. Only now I realize that "helix" in its name might have been a bit of a hint… The shell of the vast amount of these snails is wound right-handedly. Only one in about a million snails supposedly has a left-handed shell. German folklore labels such a special creature as a *schneckenkönig*, a "king of snails". For many years, I searched for my snail king; I have looked at many, many snail shells... But how do you search for left-handed snails? In fact, our hands are ideal tools to determine the winding direction

2

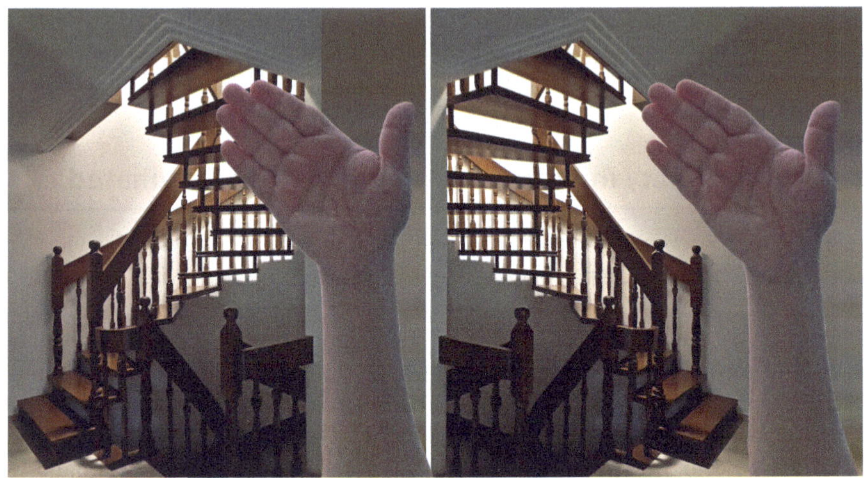

■ Fig. 2.11 **Right-handed DNA and how to find them**… With whatever winding object in front of you, you can determine whether that "staircase" is wound right- or left-handedly, by using your hands. Just in your mind, go in with your right hand, climbing the stairs. Does the thumb then point in the same direction as the stairs? Up? Then, a right-handed staircase/snail/DNA/helix is in front of you. If not, then try with the left hand. Left-handed! Still not? Then again try with the right hand, again. Any better? (Photo credit: Staircase, Duisburg, Germany, JWM, 2018; Hand, JWM, 2022)

of threads, stairs, snail shells or other helices. Please look at ■ Fig. 2.11 for more instructions.

Nevertheless, our body's DNA is a right-handed helix. Full stop. Some people almost celebrate putting wrongly depicted (i.e., accidentally mirrored) DNA on the spot. It happens easily. For left-handed DNA, just search for "lefthanded DNA" online. #HallOfShame. When in doubt, let your biotech logo, or whatever, been checked by an expert. Back to my quest, I have never found such a snail—a king of snails, until I heard about Jeremy. The following **meandering** is about Jeremy.

Meandering: Jeremy, the Left-Handed Snail

Jeremy truly was a king of snails. His shell was wound left-handed, not right-handed—as the shell of millions of other snails. As a lefty, he was named Jeremy, after the English politician Jeremy Corbyn. The photos show Jeremy first on the right, and then at the top.

For his extraordinary left-handed shell, Jeremy could not mate with "normal" snails of his very same species, for geometrical reasons. So, Jeremy's foster father Angus launched a campaign to find a partner for Jeremy. For quite a while, Jeremy was very popular in the press and on social media. Via the handle @leftysnail, Jeremy actively tweeted until maybe 2018—a snail with their own Twitter account … X. All this outreach turned out to be productive. At least two more left-handed snails were found as a result. Good for Jeremy. At the documented age of two full years, Jeremy died peacefully. He leaves behind 56 direct offspring. But… All of them have a right-handed snail shell.

This is not only a touching story, but also illustrates—in molecular detail—how a genetic change brings about a macroscopically visible phenotype. Jeremy was sequenced, and it turned out that he carried a mutation in a gene coding for the protein formin; the mutated version aligned actin fibers in a different way. The research on Jeremy & Co. is relevant to human embryonic development mechanisms. After all, the *Situs inversus* is a mysterious peculiarity of human anatomy. There, the unpaired organs of a human are mirrored. The Spanish singer and actor Enrique Iglesias is a celebrity with a well-known *Situs inversus*. All in all, an insightful demonstration that we still do not know everything,

2

even about our own earliest development.

■ **For further reading**

Not many snails have their own Wikipedia page ▸ https://en.wikipedia.org/wiki/Jeremy_(snail)

A great overview article, written by Jeremy's foster father—Davison. **2020**. Flipping Shells! Unwinding LR Asymmetry in Mirror-Image Molluscs. *Trends Genet*. 36(3): 189–202. Review. ▸ https://doi.org/10.1016/j.tig.2019.12.003.

(Image credit: Jeremy and another snail, Angus Davison, 2020; reproduction with kind permission)

But what was the third reason for homochirality again, the one that Jeremy's story conveys? I try it this way: Maybe it is the idea that the mutated formin protein, via differentially arranged actin fibers, produces a different pattern of biomineralization. In other words, one type of homochiral biopolymer may influence another. This argument follows the thinking of evolutionary scientist Koji Tamura. He thinks that one homopolymer had followed the other. First there were homochiral transfer RNAs. And they then "liked" one type of amino acids better than another. And from these "more preferred" amino acids, homochiral proteins finally evolved. We don't know in which order the different chiral biomolecules had formed; I however sincerely hope you liked the story of Jeremy.

We now know that snails with left-handed shells can be extremely interesting. But for the application-oriented reader, is there anything more about chirality and how it may be exploited biotechnologically or pharmacologically. Mixing experiments involving different proportions of D- and L-amino acids, showed that molecules are completely functional, if they were EITHER made completely from D-amino acids OR made completely of L-forms; mixing D- and L-building blocks together though was less advantageous.

The D-forms are in no way inferior to the biological L-proteins in terms of stability or activity. However, proteins from D amino acids have a great advantage. They are significantly more stable in biological systems. Theoretically, if we were to consume a D-protein shake (or a D-steak or a D-tofu burger)—all made out of D-amino acids, instead of the naturally occurring L-ones…—our digestive system probably would be quite irritated and maybe overwhelmed. Can one convert this to our advantage? The **meandering** "Mirror, mirror on the wall, who's the best binder of them all" describes how this principle might be applied to peptide-based drug discovery.

Meandering: Mirror, Mirror on the Wall, Who's the Best Binder of Them All?

There are therapeutic proteins, therapeutic antibodies, and other binders, as well as pharmacologically active peptides, such as insulin. They all consist of L-amino acids. They all are broken down quite quickly in the body. How could the stability of peptides and proteins in the body significantly increased: to expand their medical and biotechnological use? Display methods, explained more in ▶ Chapter. 8, such as phage display or ribosome display, enable us to select any L-peptides, that could bind to any protein target; normally all molecules made of L-amino acids.

The Peter Kim had a bright idea in this regard: What if the target molecule was presented mirrored as a D-protein? Then an L-binding peptide would be selected. This binder, in turn, mirrored as a D-protein, should bind the natural L-protein. At the same time, it would have a much higher physiological durability, as our body doesn't know how to break down D-amino acids. Sounds bonkers, but… it works perfectly well.

Just as a reminder, using various display methods, we can generate binding molecules like antibodies against any target molecule, not just peptides and proteins. So, what if the target molecule is not built with L-amino acids or D-sugars, but with their mirror images, the D-amino acids, or L-sugars? Once the actual selection procedure for the binding molecule is over, the resulting binding molecule just needs to be synthesized as a mirror image itself. The result is a molecular binder that is no longer recognized by the body and should therefore be much more stable in the body.

- **enCapsulated**

The more of life we can mirror, the easier we can create "bio-orthogonal" mirror images of biomolecules; thus, enabling a very special way of drug discovery. Yes, one might make money with chirality.

- **Extended Readings**

The very first paper on mirror-image phage display—Schumacher et al. **1996**. Identification of D-peptide ligands through mirror-image phage display. Science. 271(5257): 1854-7. ▶ https://doi.org/10.1126/science.271.5257.1854.

A 2021 review asks: How far have we come in mirroring all of life?—Rohden et al. **2021**. Through the looking glass: milestones on the road towards mirroring life. *Trends Biochem Sci*. 46(11): 931-943. ▶ https://doi.org/10.1016/j.tibs.2021.06.006.

Converted For mirror image strategies to succeed, the **chemical synthesis** of proteins had to improve first. Nowadays, proteins with 100 amino acids are no longer as challenging as they once were. Cleverly linking prefabricated peptide fragments, makes it possible to produce ever longer peptide chains. But why should we stop at the stage of proteins? RNA and DNA binders work as well; for fans: they are known as aptamers. They too could be optimized through mirror image phage display. So, demand has grown to synthesize nucleic acids with L-ribose, instead of D-ribose, and to be able to sequence them.

▪ Navigation

Latest insights on the subject—Deng et al. **2024**. Symmetry breaking and chiral amplification in prebiotic ligation reactions. *Nature*. 626(8001):1019-1024. ▶ https://doi.org/10.1038/s41586-024-07059-y.

2

Computer simulations highlighted homochiral polymers, be it only D or only L, to be more stable than mixtures over the two—Skolnick et al. **2019**. On the possible origin of protein homochirality, structure, and biochemical function. *PNAS*. 116(52): 26571–9. ▶ https://doi.org/10.1073/pnas.1908241116.

A homochiral biopolymer should have influenced the next one. First, there were probably homopolymer tRNAs, which then preferred L-amino acids—Tamura. **2019**. Perspectives on the Origin of Biological Homochirality on Earth. *J Mol Evol*. 87(4–6): 143–146. ▶ https://doi.org/10.1007/s00239-019-09897-1.

Here are **two** chemical instances of the **total synthesis** of the Ras protein, one from Dortmund—Becker et al. **2003**. Total chemical synthesis of a functional interacting protein pair: the protooncogene H-Ras and the Ras-binding domain of its effector c-Raf1. *PNAS* 100(9): 5075-80. 10.1073/pnas.0831227100—**and** the mirror image from New York—Levinson et al. **2017**. Total Chemical Synthesis and Folding of All-l and All-d Variants of Oncogenic KRas(G12V). *J Am Chem Soc*. 139(22): 7632-9. ▶ https://doi.org/10.1021/jacs.7b02988.

Extending mirror-image approaches to bacterial target molecules and novel D-peptide antibiotics—Adaligil et al. **2019**. Discovery of Peptide Antibiotics Composed of D-Amino Acids. *ACS Chem Biol*. 14(7):1498-1506. ▶ https://doi.org/10.1021/acschembio.9b00234.

The story is already moving on to the next round. Scientists have created the mirror image of a very precise DNA polymerase. This could be used to produce mirror-image DNA on a large scale—Fan et al. **2021**. Bioorthogonal information storage in L-DNA with a high-fidelity mirror-image Pfu DNA polymerase. *Nat Biotechnol*. ▶ https://doi.org/10.1038/s41587-021-00969-6.

When working with normal D-DNA aptamers and mirror-image L-DNA aptamers, one probably should also be able to sequence L-DNA—Liu & Zhu. **2018**. Sequencing Mirror-Image DNA Chemically. *Cell Chem Biol*. 25(9): 1151-6. ▶ https://doi.org/10.1016/j.chembiol.2018.06.005.

DNA or the Book of Life

Contents

© The Author(s), under exclusive license to Springer-Verlag GmbH, DE, part of Springer Nature 2026
J. W. Mueller, *Ultimately Understanding Biochemistry*, https://doi.org/10.1007/978-3-662-71889-6_3

A lot has been written about DNA. Much of it can be found in genetics textbooks. Here we want to discuss what makes DNA so special. We want to look at the proportions of biomolecules, relative to each other, which are often misrepresented elsewhere. Finally, we highlight a few enzymes that are indispensable for the eternal winding-up and winding-down of the "thread of life". DNA is the first somewhat longer biopolymer that we assemble from our LEGO construction kit of life, the deoxyribonucleic acid. Most likely, DNA is the longest biopolymer on the planet.

3.1 The Structure of DNA Explains Its Function

The name **D**eoxyribo- **N**ucleic **A**cid directly describes the structure of the DNA single strand: The C5 sugar **D**eoxyribose lacks a hydroxyl group at the 2′ carbon atom. Together with a **N**ucleobase, this sugar forms a nucleoside. We discussed nucleosides and nucleotides in ► Chapter 2. This nucleoside unit then forms a very special double-ester bond with phosphoric *A*cid, as the third negative charge of phosphoric acid hovers above the double ester and prevents rapid cleavage by water. If this is not yet something extraordinary, the magic begins at the latest when a DNA single strand meets a corresponding and complementary strand of DNA of a certain length. The two form an antiparallel double strand, instantly—a beautiful right-handed double helix. How the 3D structure of that helix was elucidated, is sketched in the following **meandering**.

Meandering: The Race About Solving the Structure of DNA—Part 2

In the 1930s, DNA was known as a biomolecule, but without function. Oswald Avery provided solid evidence that DNA was the molecule of inheritance, in 1943. And this was strikingly confirmed by Alfred Hershey and Martha Chase, in 1952.

The!!! Molecule of Inheritance DNA had turned into something quite some researchers wanted to decipher in its structure. So, a race around elucidating the three-dimensional structure of DNA kicked off. Maurice Wilkins described the spirit of these times, by referring to DNA as the gold of King Midas: "*Everyone who worked on it went mad*" (1).

James Watson also gives his personal account about those days—one of my favorite books, a very enjoyable read, though probably not particularly objective—"The Double Helix" (2). Watson conveys the impression that science almost exclusively takes place in pubs and at conferences.

A surprisingly "recent" source about that time is a collection of letters, kept by Francis Crick. They had been long considered lost (3). These letters evidence that Crick and Wilkins had a very friendly relationship, despite their very different characters.

Whatever happened or not happened between all these people, in presence or in absence of any other person, is best described, and speculated about, in the writing available elsewhere.

Fact is, on April 25, 1953, three papers were published—Watson & Crick had their paper, Wilkins and his team had theirs, and Rosaline Franklin published as well, together with her doctoral student, all independent publications in the same issue of the journal Nature. These days we probably would call this back-to-back-to-back publishing. Franklin died in 1958. In 1962, Crick, Watson and Wilkins were at some sort of award ceremony in Stockholm. A lot has been written about the discovery of DNA somewhere else, so we want to stop here for now. In ▶ Chapter 9, however, we will return to the actual three original publications, to comment on.

1. Maurice Wilkins—quoted from Judson HF. **1996**. The Eight Day of Creation, CSHL Press.
2. James Watson. **1968**. The Double Helix, A personal account of the discovery of the structure of DNA. Atheneum Press.
3. Gann & Witkowski. **2010**. The lost correspondence of Francis Crick. Nature. 467(7315): 519ff.

3.2 DNA Is Huge and Chemically Very Stable, the Ideal Blueprint of Life

Without DNA to guide them, our cells would not know how to correctly build proteins. The DNA in our body consists of only four letters—the nucleobases A, C, G, and T. To encode maybe 20 different amino acids, or so, words of always three bases are formed, and these triplet words are known as codons. Codon "words" have a safety copy on the other DNA strand, and it is this safety copy that determines which of the 20 amino acids, from which proteins are made, is incorporated into the current protein at the current position. This protein code doesn't know images or spaces. Four dedicated codons represent all punctuation available. There is a sign for Start, and one for Stop, and another one for Stop, … … … and—surprise—a third codon for Stop. When it comes to stopping translation, seemingly one is better safe than sorry.

Three base pairs represent a codon—the assembly instruction for a single amino acid. Probably worth looking at the molecular sizes of a codon and an average amino acid. You might think of a neatly folded IKEA instruction leaflet and some heavy flat-packed furniture, made from laminated chipboard. The calculation on **napkin VI** shows the opposite—DNA is huge compared to the encoded proteins. Less like leaflets or those yellow sticky notes, DNA and the information contained in it can better be compared to plates of stone with Babylonian cuneiform. Schematic illustrations in many textbooks often ignored the true size ratios between DNA and protein.

Napkin VI: How Heavy Is DNA?

The average base pair, A-T or G-C, weighs **660 Da***

A codon of **3** base pairs therefore weighs **1980 Da**

A codon stands for an amino acid with an average molecular weight of **110 Da**

The one divided by the other, makes us realize that the blueprint is **18 times heavier** than what is then incorporated into a peptide.

Please remember this the next time you shop at an IKEA or similar furniture store near to you.

*If you wonder what a Dalton actually is... Please look it up in the glossary at the end of this book.

If you think of DNA as something like cuneiform writing, carved in stone, you could easily think of RNA as a traced copy of it—somehow been rubbed-off. In a somewhat related analogy, let's look at British coins (◘ Fig. 3.1). DNA is large and robust. However, to make it last almost forever, we need to apply several chemical tricks. It is these decisive chemical differences that make DNA so stable, compared to RNA.

First, DNA avoids using ribose as its standard sugar backbone. That 2′-OH group of ribose would only too easily attack the phosphodiester bond. Thus, RNA is always in danger of chemically degrading itself. Using **2′-deoxyribose** in the DNA backbone instead, this reaction is inhibited, as the "dangerous" 2′-OH group is removed. This one trick already increases DNA's stability in aqueous solution, many times over the stability of RNA.

Secondly, a different nucleobase is used: 5-methyl-uracil, which we call **thymine**, is used instead of uracil. This trick is not directly understandable, as this modifica-

◘ **Fig. 3.1** **RNA is a carbon copy of DNA**. How to best describe the connection between RNA and DNA. Maybe this way: If you look at the UK penny coins that have been in circulation since 2008, you'll see some sort of puzzle pieces on their back. Only, if I assemble these coins properly and rub them off on paper, using a pencil, making a rubber copy—the final "RNA"—it becomes clear that the UK penny coins altogether represent the Royal coat of arms. Fun fact: Within this analogy, my pencil would correspond to the RNA polymerase enzyme. (Photos: Coins, JWM 2022)

tion only indirectly offers the opportunity to correct a "small" chemical weakness of ANOTHER base. It is cytosine's labile amino group at the fourth position of its ring, that is all too eager to dissolve in water, to hydrolyze, to convert to uracil. If uracil were a normal DNA base, a permanent mutation in the DNA would have already occurred. Thymine belongs into DNA; uracil should not be in there at all. This allows repair enzymes to recognize uracil as "foreign", and to cut it out of DNA and replace it with cytosine. This happens many, many times a day, in our DNA. A third trick of DNA, compared to RNA, is its 'social nature'—it is found alone only very rarely. Most often, DNA makes an appearance as a **double strand**.

Core DNA is quite large compared to other biomolecules. It is mechanically relatively robust. Because of a few chemical tricks, DNA lasts almost forever. All these properties make DNA the ideal material for the "Book of Life".

- **Navigation**

You can download folding templates for paper models of DNA from the protein data base, the PDB. When folding these DNA folding sheets, you might realize soon that folding DNA, even as paper model, is not that easy ▶ https://pdb101.rcsb.org/learn/paper-models/dna. ▶ https://doi.org/10.2210/rcsb_pdb/mom_2001_11 (Accessed on 14.4.2024)

It is not just 20 amino acids that ALL proteins are made of. Some sort of specific stop signal are sometimes used to smuggle-in the special amino acids selenocysteine and pyrrolysine into proteins, a small extension in the genetic code, which we will discuss in more detail in ▶ Sect. 3.6.

We have discussed the conversion of uracil to cytosine, briefly. But what consequences the methylation of cytosine within DNA might have, is discussed in ▶ Sect. 8.2.

3.3 DNA Does Not Always Look the Same

Watson and Crick presented B-DNA to the world, as the very idea, the essence of DNA, nearly a Platonic ideal. B-DNA is the top model of DNA. We all love it. B-DNA can be found featuring on all sorts of logos of bio-molecular research institutes or universities, on business cards of biotech companies everywhere around the world. Try searching for images, when typing in "DNA Logo" or so, into a common search engine. But how real is B-DNA really? How much does it represent everyday DNA?

Here we will look at DNA structures that deviate from this ideal image. Yes, AT-rich B-DNA might bend. Then there is A-DNA, the somewhat more laid-back version of the biomolecule. And there is the unusual, zipped-out Z-DNA. All of these are shown in ◘ Fig. 3.2. We know of G-quadruplex stacks and i-motifs. And we will mention the three-stranded DNA from mitochondria and telomeres on a napkin calculation.

Firstly, **AT-rich B-DNA sequences** can bend easily. This plays a major role at the start of transcription: "TATAA" is not a joke in the sense of "Here I am", but an

3

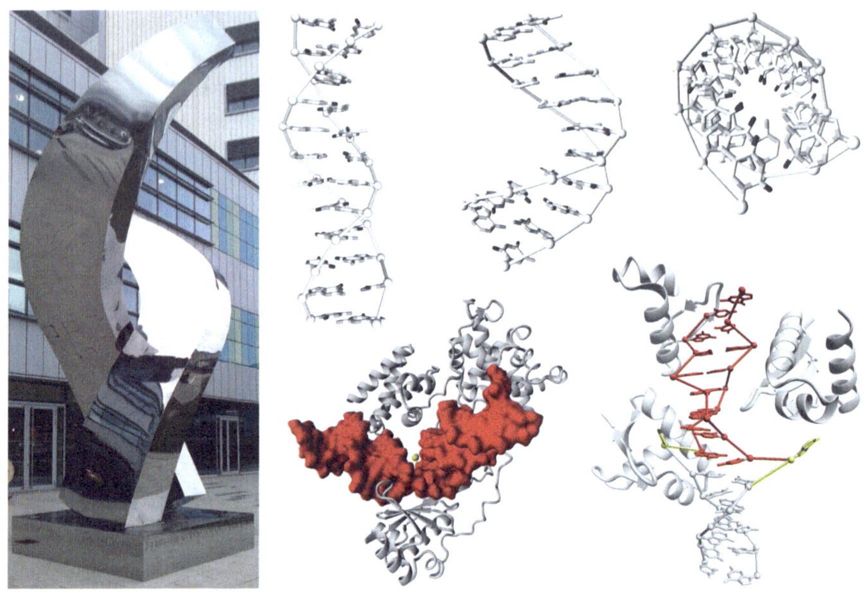

Fig. 3.2 **Some flavor of the variety of forms of DNA**. **Left**: The 8.5-meter-high DNA sculpture "To the Future", by artist Richard Thornton, prominently features at the main entrance to the Queen Elizabeth Hospital in Birmingham. **Top middle**: A short piece of beautiful and lean B-DNA. **Top right**: Two views of a piece of A-DNA, which might be considered more relaxed and settled than B-DNA. **Bottom middle**: Some AT-rich B-DNA (red)—a TATAA motif, appears quite buckled, because of binding to a DNA bending protein. **Bottom right**: At the bottom, this structure starts with boring B-DNA. But then the double strand switches into Z-DNA mode (red), which is stabilized by Z-DNA-binding proteins (partially shown). At the transition point between B- and Z-DNA, you can see a base pair that points outward awkwardly. (Image credit: DNA sculpture in Birmingham, JWM, 2016; own structure visualizations of B-DNA according to ► https://doi.org/10.2210/pdb1BNA/pdb; of A-DNA according to ► https://doi.org/10.2210/pdb1D13/pdb; of AT-rich and curved B-DNA, bound to the human protein TFIIB-related factor 2 (Brf2) according to ► https://doi.org/10.2210/pdb4ROC/pdb; as well as a crystal structure of the B/Z transition in DNA, stabilized by Z-DNA-binding proteins according to ► https://doi.org/10.2210/pdb2ACJ/pdb)

essential part of most transcription start sites. Only at a clearly recognizable TATA box kink, the transcription machinery can properly assemble, for the transcription of a gene to begin.

Secondly, DNA can adopt several conformations, other than B-DNA—the most common probably is **A-DNA**. The relaxed and somewhat spraddled A-DNA exists in plant seeds during drought and also in dried bacterial spores. It is suspected that A-DNA represents a more stable packaging form of DNA. The shortening of the double strand during the transition from B- to A-DNA is probably a driving force in the packaging of virus particles. Finally, double-stranded RNA also prefers the A-conformation—since it needs to accommodate the 2′-OH group of its ribose, which needs a bit more space than the "slimmer" 2′-deoxyribose of DNA. Please think of the A-version of double-stranded nucleic acids as the more stable, the more comfy alternative to the slim and obedient B-DNA.

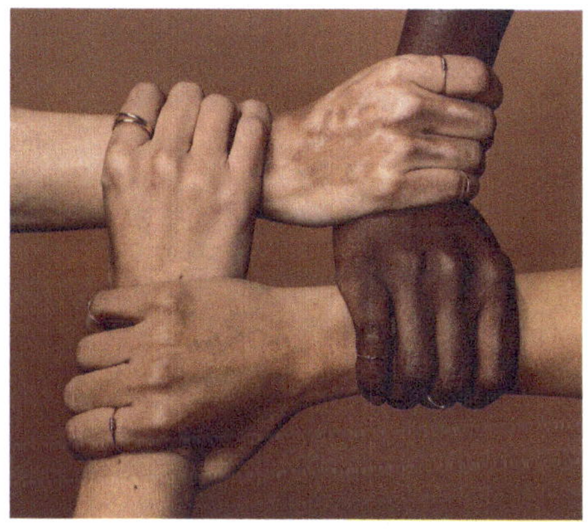

Fig. 3.3 **G-quadruplexes and the four-handed grip**. The nucleobase guanine has a "proper" Watson-Crick binding side, as the other standard bases have one. It is this side of guanine that normally binds to cytidine, DNA as usual. A certain Karst Hoogsteen first described a second type of binding in some nucleobases. Via this Hoogsteen binding site, guanine can form G-quadruplexes with itself. A G-quadruplex might be comparable to a four-handed grip. This property of G-rich DNA greatly increases the structural diversity of DNA. Interestingly, however, it seems that not all DNA researchers like G-quadruplexes. (Image credit: Four-handed grip, © Jacob Lund ► stock.adobe.com)

Totally flipped out certainly is **Z-DNA**. Short pieces of Z-DNA spontaneously form in our genome upon "negative supercoiling". They represent a way to relieve tension from extreme under-twisting of longer sections of B-DNA. Z-DNA may form, whenever normal B-DNA is unwound quickly, especially during replication, transcription, or unwrapping of histones for other reasons. A special base pair marks the transition between B- and Z-DNA, and between Z- and B-DNA, due to distortions in the DNA backbone, two bases flip far outwards. To help, there are special Z-DNA-binding proteins. It remains to be seen what other functions of Z-DNA will be discovered soon, in addition to mere tension relief.

Guanine-rich DNA can form **G-quadruplex stacks** (Fig. 3.3). If DNA is somewhat rich in the base guanine, one, or two, or even four, strands of DNA can perform a small trick. Sounds chic, doesn't it? For a start, four guanines bind to each other in a staggered fashion—the "Watson-Crick side" of one guanine base binds the "Hoogsteen side" of another G; done four times in a circle, and a layer of G's is formed. Several layers then stack together, and voila: a G-quadruplex multi-layer sandwich has formed. Pretty cool.

Hmm, have you ever heard of **triple-stranded DNA**? DNA made of three individual DNA strands, sounds absurd, completely exaggerated. Like something out of a Hollywood movie. Well, in the movie The Fifth Element (1997), an extraterrestrial life form visits Earth, and that individuum seemingly consists of triple-stranded DNA.

Unexpectedly, triple-stranded DNA is not rare at all. There is quite a lot of it in our own body. For a start, there is the **D-loop** in mitochondrial DNA. Around the replication start, an area of about 700 bases is permanently triple-stranded. This "displacement"—that's where the D-loop got its name from—makes it easier for replication to (kick) start.

This is the end. The last straw. We have arrived at the utmost ends of our chromosomes, and there is the **T-loop** of our telomeres, another place where you might spot triple-stranded DNA. Assuming that both D-loop and T-loop consist of triple-stranded DNA, **napkin VII** will tell you how much of this unusual DNA we have in us.

Napkin VII: How Much Triple-Stranded DNA Is in Us...

The D-loop of mitochondrial DNA is about **700 bp** long. In this special case, (bp) stands for both a base pair and a triplet of bases. I have to admit, that I didn't want to define a base triplet (bt), just for the calculation on this napkin.

There are about 1000 mitochondria per cell, that is then an estimated ≈ **700,000 bp per cell**

The telomere repeat sequence is 6 bp with about 2500 repetitions per telomere **15,000 bp**

In the diploid chromosome set, we have 46 chromosomes with four such ends each per cell, making **2,760,000 bp**

The total amount of triple-stranded DNA per cell is ≈ **3,460,000 bp**

Multiplied by 10^{13} body cells, that's **quite a lot!**

Core message: A considerable amount of triple-stranded DNA is in our body. It keeps the nucleoid in the mitochondrion fit and protects our telomeres from unintentional degradation. If you like, you can calculate the length and/or weight of all this DNA.

In a blink of an eye: The formation of triple-stranded DNA, observed in real time—Yamagata et al. **2016**. Triple Helix Formation in a Topologically Controlled DNA Nanosystem. *Chemistry* 22(16): 5494–8. ▸ https://doi.org/10.1002/chem.201505030.

Why do we actually have telomeres? Do we need them? "We" in this case are all eukaryotes and a few very special bacteria. **Telomeres** are the **protective caps** for linear chromosomes, just like the stiff sections at the end of boot- or shoelaces, known as aglets. Without this protective cap, the chromosome ends would shorten a bit with each cell division. A lot has been written on this topic; please search the wider literature for the "*end replication problem*". Open chromosome ends would be easy targets for somewhat aggressive exonucleases. These voracious enzymes would love to nibble on such lose strings. Repair enzymes also would be interested, they however would rather fuse two chromosomes than leave any open end exposed.

The underlying question probably is: Why do we have linear Chromosomes? After all, most bacteria and archaea lived happily and in peace, on our Earth for about 2 billion years— all with their cute little circular chromosomes. But then eukaryotes came along. Why did they have to switch to these strange linear chromosomes, which were significantly less stable and so troublesome when it comes to duplicating DNA. Linear chromosomes must have offered at least some advantages.

Well, as a start, linear chromosomes make the **winding and unwinding of DNA** significantly less problematic. Both over- and under-twisting can be compensated easily, at the rotating free ends. This property becomes particularly important because of another innovation of eukaryotes—their DNA "hair roller" proteins—the histones. As a thought experiment, try winding up a plasmid with histones—should you have one histone core per 150-ish base pairs, for a 6.3-kilobases-plasmid, this makes 42 histone cores. Don't say you knew the answer, just try to untangle the mess again, and enjoy the challenge. We probably can safely assume that circular DNA has its limit here, at some point. Linear DNA, however, can grow obscenely large. Our own chromosome number 1, for example, comprises almost 250 million base pairs. Greater **storage capacity**, therefore, could be an advantage of linear chromosomes.

When it comes to **sex**, linear chromosomes might have even more benefits. Well, here we are just talking about sexual recombination of genetic material. Meiosis—that process needed to produce sperm and egg cells from stem cells—can happen much easier with linear chromosomes. During meiosis, chromosomes of paternal and maternal origin are well mixed—think of shuffling playing cards. In addition, parts of chromosomes are exchanged in so-called cross-over events—as if you not only shuffle cards, but also cut off a part of one card and stick it to another card. Don't get me wrong, bacteria recombine their genetic information as well... They just do so somehow differently.

Then we have to acknowledge the confusing fact that not all bacteria have circular DNA. Linear chromosomes can be found here and there, for example in the bacterial species *Borrelia burgdorferi* and in different Streptomyces species. At least there, in Streptomyces, we think to know that plasmids with the ability to linearize, had integrated into the circular chromosomes. One thing became clear by looking at these bacteria: it is NOT really clear what the better setup is—linear or circular chromosomes.

What do telomeres really look like? It is not easy to correctly describe the structure of our telomeres, because two academic worlds collide over those end points. Let's start in the **T-loop** world. Dr. Titia de Lange provided the first electron microscopic images of telomeres in 1999—real photos that show a substantial loop at the end of our chromosomes—then called the T-loop. Seeing is believing, you can see T-loops, so they are real, aren't they? In addition, they come with all kinds of T-loop-stabilizing proteins that we will not discuss here.

On the other hand, there is this "loopy" enzyme telomerase, a very strange version of a DNA polymerase. Telomerase does not copy a DNA template, but a piece of RNA, that is part of the enzyme itself. Telomerase pushes this small piece forward again and again. Somewhat boringly, telomerase writes—at least in mammals—the DNA sequence TTAGGG again and again, in a long 3′-overhang, easily hundreds of copies. A bit like Bart Simpson, who has to write a sentence on the blackboard all over again. The point with the TTAGGG sequence is, that this represents a fantastic motif to form **G-quadruplexes**; especially when the DNA is a single strand only. Well, it happens that the telomere 3′-overhang is a long single strand. G-stacks as they were known for a while, were rather dead ends, knots, in the DNA topology. Newer findings around so-called parallel G-quadruplexes have removed this topological problem, as these are nicely stackable.

So, what is there at our telomeric ends of chromosomes: Are there G-quadruplexes? Or T-loops? Or both somehow? It seems that scientists who work on the T-loop like T-loops. G-quadruplex researchers love G-quadruplexes. There is a divide between the camps with hardly any overlap. If you like G-quadruplexes, then you are probably not a fan of triple-stranded DNA.

Perhaps a bit more integrative, one could say that G-quadruplexes are some sort of unique DNA structure. An ordinary DNA polymerase enzyme might choke on them. So, there are quite a few proteins whose job it is to prevent the formation of G-stacks within the cell. An interesting idea comes from another part of the chromosome—G-stacks could be a pattern for molecular recognition at the beginning of DNA replication.

Capped So, we have linear chromosomes that are protected at their ends by telomere caps. These telomeres possess guanine-rich sequences that can form G-quadruplexes. In the cell, however, telomeres are probably more likely to exist as T-loops. One thing is clear though: Telomere shortening, however, has been associated with biological ageing—if you have your telomeres under control, you don't need to worry about them getting shortened in each cell division. Seemingly, only the short telomers die young.

▪ Accompanying Literature

Mrs. de Lange provides the first photos of **telomere loops**. Back then, this really was a big deal—Griffith et al. **1999**. Mammalian telomeres end in a large duplex loop. *Cell*. 97(4): 503–14. ▸ https://doi.org/10.1016/s0092-8674(00)80760-6.

Mr. Neidle describes parallel **G-quadruplexes**, for the first time. From now on we know that G-stacks can be stacked on top of each other—Parkinson et al. **2002**. Crystal structure of parallel quadruplexes from human telomeric DNA. *Nature*. 417(6891): 876–80. ▸ https://doi.org/10.1038/nature755.

Just one example of the telomere dispute. A review article from the year 2020 cites Mrs. de Lange 27 times; at the same time it does not mention G-quadruplexes at all—Smith et al. **2020**. Structural biology of telomeres and telomerase. *Cell Mol Life Sci*. Review. ▸ https://doi.org/10.1007/s00018-019-03369-x.

An interesting idea about G-quadruplexes—they could play a role in the initiation of DNA replication—Prioleau. **2017**. G-Quadruplexes and DNA Replication Origins. *Adv Exp Med Biol*. 1042: 273–86. Review. ▸ https://doi.org/10.1007/978--981-10-6955-0_13.

Confusing, some bacteria have linear chromosomes, but this doesn't seem to suit them particularly well—Galperin. **2007**. Linear chromosomes in bacteria: no straight edge advantage? *Environ Microbiol*. 9(6): 1357–1362. ▸ https://doi.org/10.1111/j.1462-2920.2007.01328.x.

A moment of truth: B-DNA transforms into Z-DNA and then back into B-DNA—Sinden. **2005**. Molecular biology: DNA twists and flips. *Nature*. **2005**; 437(7062): 1097–8. Comment. ▸ https://doi.org/10.1038/4371097a.

3.4 Wellness for Your DNA—Gene Clippers and Conditioner

How many individual hairs do we have? I probably had a bit more than 100,000 strands on my head, a few years ago. At about 5 cm length, that would be 5000 meters of hair fiber. For Rapunzel from the fairy tale, it would be 100 times as much, so we would end up with 500 kilometers of hair. Sounds very looong, right?... Actually, it isn't that long, when you think of the thousands and thousands of kilometers of DNA that are in all our body cells, combined (**napkin VIII**).

Napkin VIII: How Long Is Our Genome, Actually?

DNA in its B form is 3.4 Ångström* long per base pair (bp). That is **3.4×10^{-10} meter**

The haploid human genome (2n) consists of 3 billion base pairs **3×10^{9} bp**

3.4×10^{-10} m multiplied by 3×10^{9} (let's cancel out a few powers) is 0.34×3 m ≈ **1.02 m**

Our haploid genome (2n) is 1 meter long. The genome of a diploid cell (4n) then, is about **2 meters**

… and this is the thread of DNA of a **single cell** only. The human body consists of about **10^{13} cells**. The unbelievable length of all our DNA threads explains the need for the existence of chromatin and all sorts of DNA-massaging proteins. Otherwise, the cellular nucleus would be a hopeless mess. If we would stretch out all our DNA linearly, from all these cells, one after the other, all that DNA would reach about **67 times from the Earth to the Sun and back**. Can you re-trace my calculation?

*If you want to know more about what an Ångström is, please check out the glossary of this book.

It is obvious that we need to take care of our hair-thin DNA; and DNA actually is much thinner than a single hair. In this case, "we" includes all living cells, that contain all the machinery to keep our DNA in shape. A great invention of eukaryotes, us humans included, were histone proteins, some sort of hair roller. Histones form a core and about 146 base pairs of DNA wrap around it. There you are: a nucleosome particle is made. Histone proteins have a rigid and structured part that acts as an actual hair roller. The N-terminus of the histone proteins, however, flaps out loosely from the packing. And it is mainly the information written on these histone N-termini that constitutes the scientific discipline of epigenetics.

There are great structures of the histone core with DNA wrapped around them, for example the PDB ID 1KX5, by Tim Richmond and colleagues. In this structure, 147 base pairs (×3.4 Angstrom about 500 Angstrom long) are wound around a core of histones. The two ends of the wound-up DNA piece are about 70 Angstrom apart. The curling results in about a seven-fold shortening—a DNA thread with a nucleosome here and there—the 10 nm fiber. Then a very special curler protein comes into play, the histone H1—possibly more like a hair clip. H1 connects nucleosomes with each other; the 30 nm fiber is created. After that, there are a few advanced tricks of curling, some of which are not yet fully understood. Ta-dah, the metaphase chromosome is finished—we can see it even in low-tech light microscopes.

Circular or straight Bacteria and Archaea have circular chromosomes. Eukaryotes, on the other hand, literally have linear chromosome monsters, most of which are significantly larger and longer than all the genetic material of most bacterial cells. Some bacteria have linear chromosomes. Well, Eukaryotes have circular chromosomes at least in their organelles—we think of mitochondria and chloroplasts. It feels as if we are going in circles...

Meandering: Ancient DNA Allows a Look Into the Past

DNA is a fairly stable chemical, and modern methods for amplifying and sequencing DNA have literally reached an industrial era, allowing DNA analysis from all sorts of sources. What to do with DNA then, with that killer stable molecule? Probably best to use it for **forensic DNA analysis**. The tiniest traces of skin, hair, blood or other cell-containing body fluids, left at the crime scene, can serve to narrow down the circle of suspected people. In former days, dedicated DNA regions were analyzed that showed variation both in sequence and in length between individuals of a certain population. Nowadays, it is more likely than not, that the genome of the suspect will be sequenced in its entirety.

DNA does decay eventually—so, preferably you would extract DNA from fresh specimens. Under some circumstances, “fresh” in this context still can mean hundreds or thousands of years. It is probably realistic to look back 100,000 years or possibly even half a million years. Hence, a lot of research on ancient DNA focused on humans themselves, other hominid species, and on our best friends, be it the dog or the horse. Hollywood has *slightly* overestimated the possibilities of ancient DNA from bones or blood, preserved in amber. It is safe to say: There is no 65-million-year-old dinosaur DNA that we can analyze.

The analysis of old or even ancient DNA nicely complements the already existing analytical methods of archaeology. Ancient DNA caused a complete rewriting of the history of human evolution and human migrations. Back in the twentieth century, ancient DNA also helped to resolve inheritance disputes. Disney *reported* about the case of Anastasia.

The **illustration** in this meandering shows Princess Anastasia, the daughter of the last Tsar of Russia, the skeleton of an Allosaurus named Rowry from the Lapworth Museum of Geology (University of Birmingham) and a reconstruction of a Neanderthal.

Another bit of ancient DNA is included in ▶ Chapter 8, the chapter on evolution. There we discuss when humans really went to the dogs.

(Image credits: **Left**: Photo of Anastasia, the Russian Tsar's princess Anastasia Nikolajewna Romanowa. Reproduction with kind permission of the Archive of the Library of Congress of the United States of America; **Right top**: Skeleton of a carnivorous Allosaurus dinosaur. Taken on a "Lapworth Lates" Night of Science outreach event at the Lapworth Museum of Geology, University of Birmingham. The Allosaurus is nicknamed Rowry. Photo, JWM, 2020. **Right bottom**: Illustration of a Neanderthal, © Federico Gambarini / dpa / picture alliance)

3.5 The Universal Code—The Alphabet of Life

Every living creature uses the same alphabet to translate DNA into amino acids. The very few exceptions are limited to a specific cell organelle here or there and a few very special species. One clearly can classify these exceptions-to-a-universal-rule as secondary developments. Is the universal code simply another evolutionarily fixed coincidence ("frozen accident")? Or was there some sort of optimization at play in the transition from the pre-biotic world to the DNA-encoded life?

Let's look at the actual code more closely. Some form of representation of the code should readily be available for you—be it as a table or as a radial representation that is referred to as "codon sun" in German textbooks. Let's start with what happens, if one codon "word" would be perturbed. A single base pair might be exchanged—a realistic scenario for a DNA mutation. What happens upon a point mutation? Quite often nothing happens. That's mostly when the mutation hits the third position of the codon. For a lot of other codon changes, a lot of mutations, don't do nothing, but they cause surprisingly little damage. The universal code isn't random –amino acids with similar properties are usually coded by similar codons. All of this suggests that the universal code is an optimized code that neutralizes or at least mitigates many, but not all, mutations.

Milton Saier provided a fresh look at the universal code in 2019 (I provide an adapted version of his scheme in ◘ Fig. 3.4). Saier impressively shows that the code is read at the middle position first. The middle position determines the general type of amino acid. Only then, the first position is decoded. It specifies the actual amino acid. And lastly, there is the third position. At times, this position is completely irrelevant. In quite some instances, it only matters if it is a purine (A or G) or a pyrimidine base (C or U). The actual nucleobase in the third position is

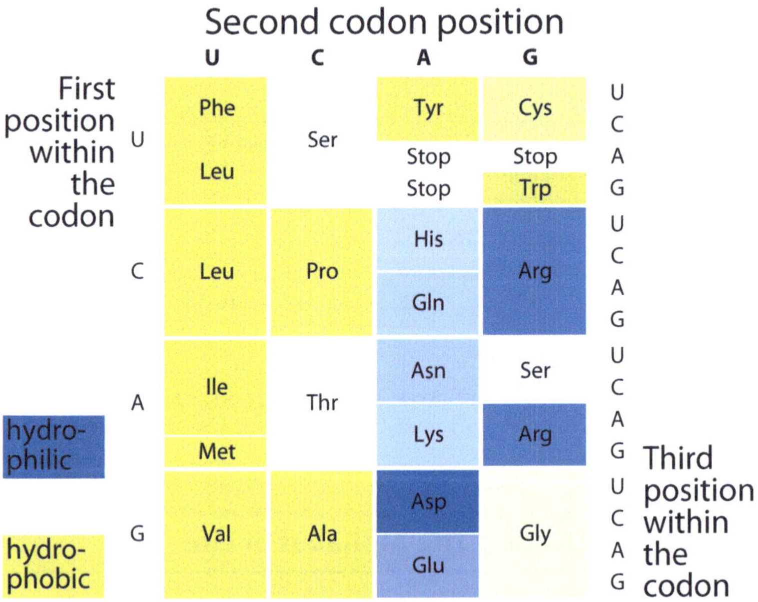

Second codon position

First position within the codon	U	C	A	G	Third position within the codon
U	Phe Leu	Ser	Tyr Stop Stop	Cys Stop Trp	U C A G
C	Leu	Pro	His Gln	Arg	U C A G
A	Ile Met	Thr	Asn Lys	Ser Arg	U C A G
G	Val	Ala	Asp Glu	Gly	U C A G

hydro-philic

hydro-phobic

◘ **Fig. 3.4** **In the universal code, a lot is determined by the second nucleobase**. Here is a representation of the genetic code, sorted by the second nucleobase, as suggested by Dr. Saier in 2019. The hydrophobicity of the side chains of all 20 amino acids that normally occur in proteins was taken from the study by Zhu et al. (**2016**). These values were color-coded between yellow and blue and applied to the codon table. A clever way of measuring hydrophobicity of amino acid side chains—Zhu et al. **2016**. Characterizing hydrophobicity of amino acid side chains in a protein environment via measuring contact angle of a water nanodroplet on planar peptide network. *PNAS*; 113(46):12946–12,951. ▸ https://doi.org/10.1073/pnas.1616138113. (Illustrations: Code table, JWM, 2025) important in two instances only—when coding for start and methionine, and when coding for stop or for tryptophan. As an example, if the middle position of your codon is "U", it can only become a hydrophobic amino acid, no matter what you put before or after it.

With this new way of interpreting the genetic code, Saier offers a plausible way of how the code might have developed. Initially, there could have been a one-letter code for very simple proteins, consisting of four building blocks only. That code then was expanded to a two-letter code, allowing for incorporating more variation. Ultimately, the three-letter code was established. This stepwise extension explains why closely related amino acids so often are coded by similar codons. And as a secondary outcome, the resulting code is less susceptible to mutations. A win-win situation.

Coded Representations of the genetic code are so commonly found in textbooks that we easily forget that the deciphering of the universal code was a scientific race of its own. A **meandering** in three acts describes that race. After all, the genetic code is a code without a period or comma. The many nucleotides only make sense when a **reading frame** is established. A dedicated start-tRNA, which happens to be loaded with the amino acid methionine defines the reading frame. This reading grid is robotically applied then to the rest of the mRNA, be it 50–70 amino acids, or 2000–5000 amino acids. We are not done with the genetic code—**Napkin IX** concludes this chapter with a few considerations on reading frames.

Meandering: Moscow and the Deciphering of the Genetic Code—A Story in Three Acts

First Act: The structure of the DNA double helix immediately explains how DNA replication might work. Nonetheless, how proteins are formed from DNA remained elusive. So, it happened that some reputable and ambitious gentlemen, including James Watson and Francis Crick, founded the "RNA Tie Club". The 20 members named themselves after the individual amino acids—Watsons cover name was proline, and Cricks name was tyrosine. The club met once or twice a year for scientific exchange. In the end, it was someone from outside the club who made the great discovery regarding the genetic code.

Second Act: Heinz-Günter Wittmann was working on deciphering the genetic code in Tübingen, around 1960. He was working on chemically induced mutants of the tobacco mosaic virus. Most of the variants he laboriously analyzed showed no changes in amino acid composition. There were some mutants with a single different amino acid. Very few of his mutants contained two amino acids changes. Wittmann tried to correlate the observed changes in the amino acid composition with the theoretically possible RNA mutations. Wittmann presented his data at an *ad hoc* colloquium organized by Crick, all at the fifth International Biochemistry Congress 1961 in Moscow.

Final Act: Marshall Nirenberg was a newly appointed NIH group leader in Bethesda, Maryland, near Washington, D.C. Together with his German postdoc

3

Heinrich Matthaei from Bonn, Germany, he developed a cell-free translation system. The *eureka* moment was when they added RNA that consisted only of uracil letters to their cell-free translation system—it produced a chain of phenylalanine residues. "UUU" was thus the first deciphered codon. This system was much easier to handle than Wittmann's virus variants. More codons followed, word for word. In 1961 in Moscow—at the same *ad hoc* colloquium—Nirenberg was celebrated for his work.

Later, Nirenberg received many awards. Wittmann turned to other topics. The contribution of Heinrich Matthaei was unrightfully omitted and forgotten. He never really recovered from this.

For the fact check: The RNA Tie Club never acted particularly publicly. Therefore, here is a link to the chapter "The RNA Tie Club JDW/2/7/2" of the James D. Watson collection of the Wellcome Trust's archive ► https://wellcomecollection.org/works/jwyc-znfk—**and** to the English Wikipedia page—► https://en.wikipedia.org/wiki/RNA_Tie_Club (read on 22nd February, 2025)

Hans-Jörg Rheinberger looks at the German part of the history of the universal code—Rheinberger. **2018**. Heinz-Günter Wittmann—a pioneer of the genetic code. *BioSpektrum*. ► https://doi.org/10.1007/s12268-018-0989-3.

A 2009 interview of Dr. Nirenberg, where he does mention his co-worker Matthaei—Ginsberg. **2009**. Deciphering the Genetic Code. A National Historic Chemical Landmark. ► https://www.acs.org/content/acs/en/education/whatischemistry/landmarks/geneticcode.html (read on April 16, 2022)

Napkin IX: What's so Special About an Open Reading Frame?

The short answer is, it wouldn't exist by chance. Let's take a completely random sequence. If there was an ATG start codon somewhere, opening a reading frame, how quickly would this reading frame be closed again by one of the three stop codons?

1 codon out of 64 → 1.5% probability for methionine

It continues if no stop codon comes, so with 61 out of 64 codons → **that's about 95%**

The same probability applies to each new codon, which is multiplied by all of the above.

Finally, the reading frame must have an end as well → **4.7% for one of the three stop codons**.

The probabilities for *methionine* × *(probability for everything except stop)*^{n} × *probability for any stop codon* are multiplied in the end, where *n* corresponds to the number of amino acids. Don't forget to take the N-terminal methionine into account.

For such a small gene like insulin (110 amino acids including pre-peptide and C-peptide to be cut out), we arrive at **0.000 003 9%** probability.

This quite small probability is for ANY reading frame with 110 amino acids length and proper start and stop signals to occur, encoded by a random sequence—so far without any biological sense. Can you imagine how small chance get when it is about the "random" conservation of an essential gene?

Conserved: If you see a longer reading frame in a DNA sequence, it is always something special. However, if you take the very small probability for a reading frame and multiply it with the huge number of nucleotides in the one or the other genome, apparent reading frames might appear by chance, here or there.

More for fans: A new definition of an Open Reading Frame—Sieber et al. **2018**. The Definition of Open Reading Frame Revisited. *Trends Genet*. 34(3):167–170. ▸ https://doi.org/10.1016/j.tig.2017.12.009.

▪ **Further Reading**

Those who want to read more related stuff, please search for "Eugene Koonin" and "Code". Mr. Koonin has published extensively on this topic, for example this piece—Koonin & Novozhilov. **2017**. Origin and Evolution of the Universal Genetic Code. *Annu Rev Genet*. 51: 45–62. Review. ▸ https://doi.org/10.1146/annurev--genet-120116-024713.

A fresh look at the universal genetic code, sadly overlooked—Saier. **2019**. Understanding the Genetic Code. *J Bacteriol*. 201(15): e00091–19. ▸ https://doi.org/10.1128/jb.00091-19.

3.6 On Variants of The Code and Other Biological Codes

Should the genetic code turn out to be not complicated enough for you, the following question inevitably arises: Are there any other biological codes?—Of course, the answer is "Yes, but..." Before we turn to those other codes, however, we will first look at a few more features of the universal code.

The dogma—or mantra—of molecular biology is the omnipresent code that translates DNA into RNA and then into proteins—DNA—RNA—protein. That already is a bit too simple—actually, the universal code has two parts. With DNA—mRNA—tRNA—protein, the first part is already a bit more complicated than suggested initially. The other part of the universal code links to an intricate arsenal of aminoacyl-tRNA synthetases, which activate inert—lazy—amino acids and then link them to their corresponding tRNA(s), nearly always to the correct tRNAs, with astonishing precision.

Even within this conserved frame, life can be innovative. There are at least two amino acids that the ribosome directly incorporates into peptides, but only under special circumstances. In both cases, some stop codons are reinterpreted, due to unique contexts. The corresponding **meandering box** links these phenomena to semi-precious gems.

Meandering: The Ribosome Sneaks Additional Amino Acids Into Proteins, Using Opal and Amber Gems as Compensation

In almost all organisms—including humans—some "amber" UAG-stop codons are interpreted as a signal for the incorporation of selenocysteine—maybe think of it as highly redox-reactive super-cysteine. Pyrrolysine is a very strange amino acid that is built into proteins from archaea and some bacteria at very special "opal" UGA-stop codons. The exact molecular mechanisms for the incorporation of selenocysteine and pyrrolysine, would fill hours of hard-core molecular biology lectures.

Now would be a good place to explain why stop codons have been named after semi-precious gems. Scientists Rich Epstein and Charles Steinberg discovered suppressor mutants. They had a diligent student in their lab—Harris Bernstein. Instead of naming him as a co-author on the respective publications, they promised him for his efforts: From now on all UAG suppressor mutants would be called *amber mutants* in his honor—Amber (UAG), because Bernstein means *amber* in English. And, in the same vein, mutants in the other two stop codons would be named Opal (UGA) and Ochre (UAA). Isn't this terrific. These days we probably would think differently about authorship, including Mr Bernstein as an author on the actual publication.

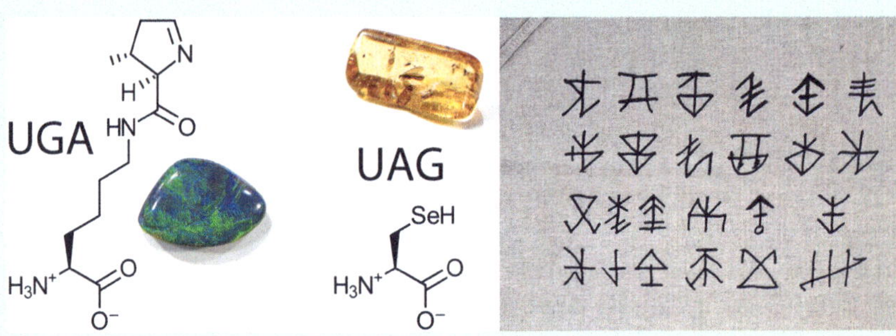

Biotechnologically, these are two systems that invited themselves to be put into use to incorporate alternative amino acids on the protein. Easy, isn't it? To re-educate an entire cell, let's think of the *E. coli* bacterium for now—to use a new core base, or two, and then build designer proteins at the ribosome, sounds "somewhat" more complicated, but seemingly only just. Synthetic biology pursues all approaches mentioned above, to reprogram the genetic code. One can only curiously wait for whatever complex natural substance these semi-synthetic approaches produce.

And now for something completely different. A vastly different code is depicted. You can see the house brands of the old fishing families on the island of Hiddensee in the Baltic Sea. With patterns mostly easy to carve, families marked their property, be it their hut or house, or their boat and buoy. Most of these signs are very old. Initially preserved in form, but later varied… Quite possibly at a time, when there were two families of the name Schmitz (one of several German versions of Smith), or five families with the name Müller on the island.

- **For further reading**

Chin. **2017**. Expanding and reprogramming the genetic code. *Nature*. 550(7674):53–60. Review. ▶ https://doi.org/10.1038/nature24031.

Hatfield & Gladyshev. **2002**. How selenium has altered our understanding of the genetic code. *Mol Cell Biol*. 22(11): 3565–76. Review. ▶ https://doi.org/10.1128/mcb.22.11.3565-3576.2002.

(Image info: Opal, © marcel ▶ stock.adobe.com; Amber with insect inclusion, © Minakryn Ruslan ▶ stock.adobe.com; Kitchen towel with the house brands of Hiddensee, JWM, 2018)

Asked more generally: What are the requirements for a biologically useful code? Certainly, some sort of suitable writing material… Then we need something to write, some sort of reading device and an eraser, to make a mark undone. In the office, these four components could be paper, pencil, the reader's eye, and the eraser or rubber. Whenever we discuss biological codes, it doesn't hurt to look back at the genetic code, the universal code, from time to time. This might help in assessing whether the system-just-looked-at really is a biological code or not.

All (bio-) **informatics** is based on a simple form of code. In binary language, the presence or absence of a "thing" is enough to express "**one**" or "**zero** ". At the molecular level, that "thing" can be as diverse as your imagination. Several "ones" and "zeros" then could be combined into more complex units. In the early days of computer science, various numbers of bits were grouped into "words". The byte was born. Somehow, the byte, built out of eight bits, prevailed in the end. In principle, one can understand how these ones and zeros were read from magnetic tapes or from rotating plastic discs. How information is stored, read and deleted in ultra-high-density USB flash memories, or in the cloud, remains astonishing for the author of this book though. In biological systems, information is rarely encoded just using ones and zeros. Often, quite complex molecular patterns represent signals. Here are three examples for biological codes:

The histone code and epigenetics: There is a lot of talk about epigenetics and the histone code. Fact is that DNA wraps around a core of histone proteins. While the C-terminal parts of these histones form quite a solid structure, their N-terminal parts reach out of the nucleosome particle; these fluffy or loose parts are presented freely to specialized enzymes to add modifications here or there, with biochemical precision. What modifications do we know? There is at least methylation (of histone proteins, or even the DNA itself), phosphorylation, acetylation, and ubiquitination. Tyrosine-sulfated histones were identified recently. A lot of additional chemistry might be going on, involving DNA and the histone tails. A matter of debate, however, is to what extent these modifications are inherited. Some signals are short-lived, they can be erased at some point. Some models suggest that environmental information could still be passed on to the offspring. Yes, you might pass on your love for biochemistry to your children. Thanks for the note.

The compartmentalization code corresponds to the idea of zip codes or address labels on proteins, to show them the way within the cell. There are all sorts of signals. They can be at the N- or at the C-terminus of a protein or somewhere in between, only to be interpreted by the complex sorting machinery of the cell. This sorting tends not to be perfect. Let's hope that at least a certain proportion of the many copies of a translated protein arrives where it is needed the most. You could see it as a numbers game. As long as a small fraction of any complicatedly folded and modified, but highly active protein ultimately makes it to the surface of a cell as a receptor, all might be alright.

The sulfation code of heparan sulfates: Proteoglycans are long chains of sugars that are attached to proteins. Often, these sugar chains consist of two alternating sugar units—that disaccharide unit then is repeated again and again. Heparan sulfates are such proteoglycans. With a certain length of the sugar chains and various attachment points on the protein scaffold, heparan sulfate might be regarded as complicated enough. Special modifications of the sugars add an additional layer of complexity to heparan sulfate.

Sulfation of the sugars at specific hydroxyl groups, certainly is the most important modification here. The intricate thing about heparan sulfation is that the involved sulfotransferases are somewhat "social" enzymes. They functionally interact with each other. Only when a sulfate group has been added "here", another one is added "there"; otherwise that whole patch remains unsulfated. This creates patterns along the heparan molecule—there are areas almost completely void of sulfates alternating with highly sulfated areas. These barcodes or fingerprints then turn into binding sites for certain signaling substances.

Cryptic: In addition to the universal code, we have discussed three examples of biological codes. There are many other codes, such as the code of hormones and signaling substances—for sending messages from one cell to another cell. There is the splicing code, the blueprint for how to make a smart ready-to-translate mRNA out of a humongously large first RNA transcript. This is how far we want to discuss codes here. Please read about code theories and other exciting biological codes somewhere else.

▪ Further Reading

We start off with something about **DNA methylation**—Aliaga et al. **2019**. Universality of the DNA methylation codes in Eucaryotes. *Sci Rep*. 9(1):173. ▶ https://doi.org/10.1038/s41598-018-37407-8.

An absolute classic, one of the "founders" of the **histone code**, Bryan Turner from Birmingham, defined the epigenetic code in 2007—Turner. **2007**. Defining an epigenetic code. *Nat Cell Biol*. 9(1): 2–6. ▶ https://doi.org/10.1038/ncb0107-2.

Excitingly, the same Dr. Turner writes in 2017 that an **epigenetic code** " *might exist, (…) but we are not there yet!* "—we haven't finally shown that this code exists! Not yet!—Turner. **2017**. Epigenetics can free us from the tyranny of selfish DNA. *The Biochemist*. 39(5): 4–7. ▶ https://doi.org/10.1042/BIO03905004.

Wow, **sulfation** joins the "club" of possible histone modifications—Yu et al. **2023**. Histone tyrosine sulfation by SULT1B1 regulates H4R3me2a and gene transcription. *Nat Chem Biol*. 19(7):855–864. ▶ https://doi.org/10.1038/s41589-023--01267-9.

Heparan sulfate, is it a code of sugars or a sulfation code?—Poulain & Yost. **2015**. Heparan sulfate proteoglycans: a sugar code for vertebrate development? *Development*. 142(20): 3456–67. Review. ▶ https://doi.org/10.1242/dev.098178.

Adding another level of complexity—as a (co-) receptor, **heparan sulfate** might be involved in all sorts of signal transmission—Hayashida et al. **2022**. Coreceptor functions of cell surface heparan sulfate proteoglycans. *Am J Physiol Cell Physiol.* 322(5): C896-C912. Review. ▶ https://doi.org/10.1152/ajpcell.00050.2022.

3.7 What Was There Before DNA? The RNA-World-Hypothesis… and What Could the Future Bring?

What came first—the chicken or the egg? This question is well known. This question even has become an umbrella term (for fans: a "hypernym")—"chicken-or-egg questions" are a certain type of question. DNA is such a great information storage, and proteins are the undisputed champions of catalysis, with RNA ranging somewhere in between. "How could all this have arisen simultaneously?"… is THE chicken-or-egg question of molecular biology. Before we try to answer this difficult question, a **meandering** follows about a very English chicken-or-egg question: What came first, the pubs on the canals or the canals at the pubs.

Meandering: Which Came First, Pubs or Canals?

The Industrial Revolution started in the West Midlands of England. From the mid-eighteenth century maybe, canals were being built all over Britain, as main routes of transport. Quite a few of these canals are still in use today. Digging out the canals with shovels was ultra hard back-breaking work. Along the canals, there were numerous pubs, to provide the workers with essential alimentation, such as simple food and beer. But what came first—the pubs or the canals? Without pubs, no canals. Without canals, no supply for pubs. Not an easy question.

These days, canals are mainly used recreationally. So, you can think about this question perfectly, while walking along the canals and stopping at one of these pubs. Of course, you can also hire a narrowboat to travel on one of the canals and then dock at one of these pubs.

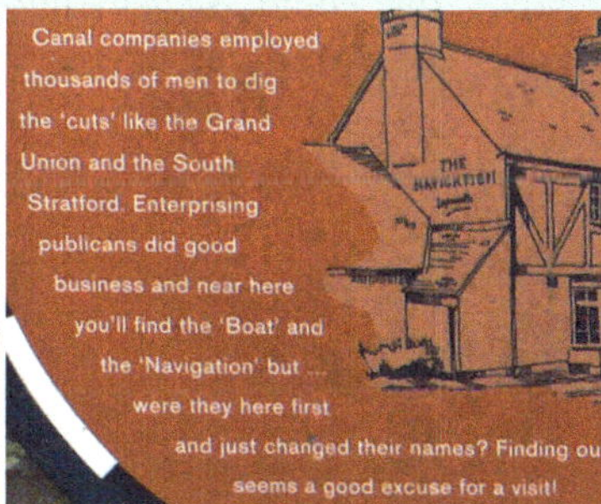

(Image information: Canal and information sign near The Navigation pub, not far from the Kingswood Junction canal cross in the southeast of Birmingham, JWM, 2016)

Back to the chicken-or-egg question at hand. For as long as life has existed, DNA was the perfect information storage and proteins were pretty good biocatalysts. When several things work hand-in-hand really, really well, it often is hard to imagine what could have existed before. But what was there before DNA? An attempt to explain is the postulation of an RNA world—a prehistoric world in which RNA was both storage medium and biocatalyst. As messenger RNA, RNA is a temporary carrier of information for proteins. Compared to DNA, RNA storage is somewhat limited in time, due to its lower chemical stability; but, yes, RNA could serve as an information storage of some kind.

And then came Tom Cech (more about him in ▶ Sect. 5.2). He discovered the first self-splicing introns. Suddenly, some kind of catalytic RNAs existed—RNAs can be enzymes. Almost all the ingredients for the RNA world were available now. The few little things still missing, such as cofactors, will be discussed in other chapters. Conceptually, this chicken-or-egg question was solved—it was no longer necessary for DNA and proteins to be present from the very beginning.

The RNA world hypothesis probably is popular, for another reason. DNA itself, more precisely how its building blocks are made in living cells, might cause headaches among prebiotic chemists. The reduction of ribonucleotides to proper **DNA building material**, the deoxyribonucleotides, is one of the central reactions of our metabolism. All known life forms afford to apply a risky radical mechanism in DNA making—nothing you'd particularly want happening in your own cells.

Could the evolutionary history of the enzymes involved provide clues about the origin of DNA-based life? A nice idea that several researchers have already explored. There seem to be three different types of these ribonucleotide reductases, and all three of them have found a different way to generate those dangerous radicals. For now, the path of scientific discovery does not continue here.

So, what does the **future** bring? DNA technology has long left the molecular biology labs of this planet. It has infiltrated many areas of research and everyday life. Forensic DNA analysis, DNA origami, and modern animal and plant breeding are just a few examples. Modified cells can produce all sorts of pharmaceutically valuable substances and in computer science, DNA itself has come into focus as a data storage medium.

For DNA, one could ask why life has settled on the core bases A, G, T, and C. Many years ago, the pragmatic answer to this question was "because there is just no other way". This answer has since been discarded. There are more and more alternatives for the backbone of deoxyribose and phosphate. Also, several new bases and base pairs have been developed that fit perfectly into the DNA double helix framework.

▪▪ Core innovation

Research into alternatives to DNA started as a scientific play-ground activity in prebiotic chemistry and theoretical biochemistry. Meanwhile, biotechnological applications are at market-ready level or at least within reach, where proteins and nucleic acids are made even better by the incorporation of alternative DNA building blocks, new nucleo-bases, coding for smarter amino acids at times.

▪ Food for Thought

DNA building blocks are made by ribonucleotide reductase. This paper explains on many pages of text that it is not that easy to determine the actual age of DNA—Lundin et al. **2015**. The origin and evolution of ribonucleotide reduction. *Life* (Basel). 5(1): 604–36. Review. ▸ https://doi.org/10.3390/life5010604.

Alternative backbones for nucleic acids are discussed here—Handal-Marquez & Pinheiro. **2021**. Life orthogonal. *The Biochemist*. 43.6: 40–43. ▸ https://doi.org/10.1042/bio_2021_191.

Anything goes. The ribosome seems to tolerate more varied mRNA building blocks than ever imagined. New letters in the genetic alphabet—Ledbetter & Romesberg. **2018**. Expanding the genetic alphabet and code. *Curr Opin Chem Biol*. 46: A1-A2. Editorial. ▸ https://doi.org/10.1016/j.cbpa.2018.09.007.

Want to read more stories about DNA—Mueller. **2022**. 11 ½ unusual facts about DNA—or what you can also do with DNA. German edition. *Springer-essential*. ▸ https://doi.org/10.1007/978-3-658-37770-0.

For somewhat dryer facts about DNA, please contact textbooks of genetics.

How Does Order Arise from Chaos?

Contents

© The Author(s), under exclusive license to Springer-Verlag GmbH, DE, part of Springer Nature 2026
J. W. Mueller, *Ultimately Understanding Biochemistry*, https://doi.org/10.1007/978-3-662-71889-6_4

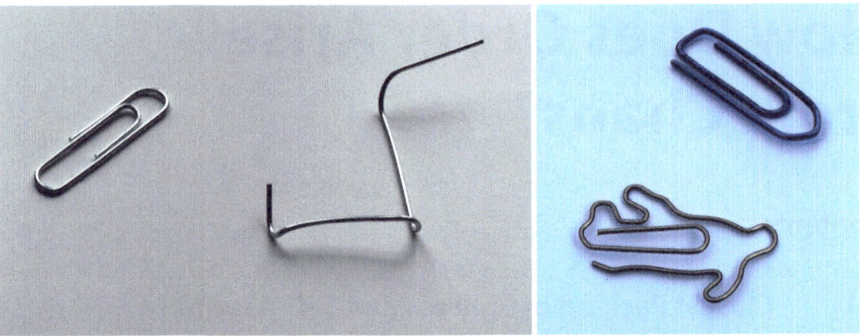

4

Fig. 4.1 **The structure of a paperclip determines its function**. A piece of wire can only be used as a paperclip if it is shaped like a paperclip—structure determines function. I saw this analogy at a lecture by crystallographer Ada Yonath in Birmingham. Good analogies are those where you can go even further within them—here I also show a paperclip from an oxidizing milieu (one that had been in the washing machine) and one that has acquired further domains. (Image credit: JWM, 2020)

Simply stitching some nucleotides together into a looong thread of DNA, was already astonishingly complicated. Endless DNA chains have their own unique problems, such as torque. Leaving the covalently linked DNA aside for now, we want to look at how some biological systems align themselves. How is it possible, for example, that some seemingly boring lipids form stable and complex biomembranes? Finally, we want to get a glimpse of understanding of how peptide chains spontaneously collapse together into highly complicated proteins. A common theme is that complicated systems can quickly emerge from relatively simple components. It is crystal clear: The structure determines the function (Fig. 4.1). But then what determines the structure?

4.1 Closest Packing of Spheres and Self-Assembly Systems

The cosmos steers towards chaos, inevitably. That at least is what we learn and teach in relation to entropy and the second law of thermodynamics. Why then is not everything around us (and including us) totally chaotic and disordered? Please not refer to our planet's dire situation at this point; let's also refrain from thinking of the current state of the desk in my office.

Apparently, some systems simply want to fall into order. Mesmerizing to see how a bee colony builds perfectly shaped hexagonal honeycombs into a suitable cave. Why don't they build five- or seven-sided combs, one might ask? **The densest packing sphere** provides a good explanation for the bee hexagons: Just put a handful of glass marbles in a flat container, then slightly tilt it and maybe rock it a little bit—these balls will mostly arrange themselves hexagonally. The round shape of the glass marbles permits this behavior, exclusively. Instead of marble balls, you can also think of a pile of oranges or apples, nicely arranged in the supermarket nearby.

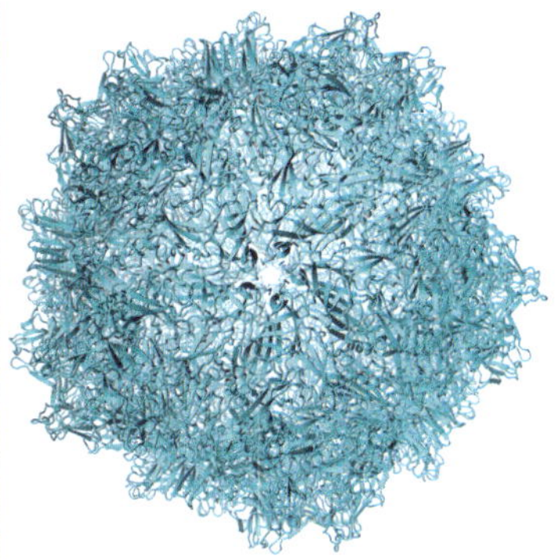

Fig. 4.2 **Giant's Causeway and a viral Capsid**. In the far north of Northern Ireland, thousands of basalt columns formed about 60 million years ago. The natural wonder is about five kilometers long, the rock layer is up to 25 meters thick. The vast majority of columns are hexagonal, but some also have four, five, seven or eight corners. A special place of legends and folklore. And then we have a small and innocent viral envelope protein, such as the VP1 protein BatAAV-10HB from an adeno-associated virus from bats. VP1 is just 58 kDa in size. At optimal chemical conditions, 60 copies of VP1 spontaneously flip together to form a molecular football. That empty virus hull now weighs about 3.5 mega-Daltons; almost the weight of the gigantic 80S ribosome. (Image information: Sunset at Giant's Causeway, © VanderWolf Images ▶ stock.adobe.com; own structure visualization according to ▶ https://doi.org/10.2210/pdb6WFU/pdbEM)

We have entered the realm of **self-assembly systems**—systems that can create order from within themselves. Giant's Causeway at the northernmost tip of Northern Ireland are the petrified remnants of a huge self-assembly system. Fig. 4.2 shows a true molecular giant, a viral capsid, next to that dam of the giant. The huge capsid is assembled from many copies of only one innocently appearing little viral protein, the VP1 protein from an adeno-associated virus from bats. When infected, cells produce the VP1 protein in large numbers. Then something weird happens—60 copies of VP1 form an icosahedron quite suddenly and spontaneously; and enclose viral nucleic acids and a few other ingredients—a virus particle is made. An icosahedron has 20 faces, all of which are beautifully evenly shaped pentagons. That small, initial VP1 envelope protein makes forming this shape possible. Biochemical self-assembly seemingly always is about hiding as many internal docking or binding sites as possible, only to adopt an as compact shape as possible—molecular cuddling. The specialty of self-assembly is that attractive forces do not necessarily need to be strong—a multitude of even the weakest biochemical interactions can cause strong cooperative effects.

Weak, but repeated, interactions not only lead to self-assembly, they could also cause the spontaneous de-mixing of homogenous mixtures of biomolecules—this then is called **liquid-liquid phase separation**. Water-oil mixtures are representatives of this category—have you made a vinegar-oil salad dressing lately... Vigorously mixed, the dressing becomes quite uniform. Forgotten in the kitchen somewhere, the mixture will eventually de-mix. Be it water or oil—one prefers to keep to themselves. A whole collection of cellular organelles without a cell membrane might not exist without phase separation, as we know them—let's think of the nucleolus, of stress granules or even of heterochromatin. Phase separation is trending, currently. New examples of phase separation are reported about every day.

Conglomerate Quite some biological systems are self-assembly systems. Order arises seemingly out of nowhere, but it actually is caused by many weak interactions. These interactions can stem from packing effects, such as close-packing of equal spheres, or by weak interactions among many copies of suitable monomers, usually favored by the hydrophobic effect. Cellular organelles without a cell membrane may form due to de-mixing, which we call phase separation.

▪ **Navigation**

Protein-biochemical craftmanship on a viral envelope protein in its isolated form, that—when treated properly—will form a viral capsid in the form a perfect icosahedron and almost the size of a ribosome—Schmidt et al. **2000**. Mechanism of assembly of recombinant murine polyomavirus-like particles, *J Virol.* 74(4): 1658–62. ▶ https://doi.org/10.1128/jvi.74.4.1658-1662.2000.

Others write review articles. These colleagues instead provide a *Leading Edge Primer* on phase separation. Well.—Sang et al. **2025**. The dynamic and heterogeneous composition of biomolecular condensates and its functional relevance. *Nat Rev Mol Cell Biol.* ▶ https://doi.org/10.1038/s41580-025-00897-2.

You can read some critical thoughts around phase separation and the hype around it from Andrea Musacchio, from Dortmund, Germany—Musacchio. **2022**. On the role of phase separation in the biogenesis of membraneless compartments. *EMBO J.* 41(5): e109952. ▶ https://doi.org/10.15252/embj.2021109952.

4.2 Biological Membranes—or—Divide and Conquer

"*Divide and Conquer*", was a motto, not only Caesar and Napoleon adhered to, when they divided larger areas (of captured land) into smaller zones to better control them. The strategy is still used these days. Let's call those small, sub-divided zones, compartments for now, at least when looking at a normal eukaryotic cell. Phase separation, mentioned just above, may create some of these compartments. It is membranes, however, that separate most of the different cellular compartments, from the rest of the cell. But what are membranes made of again?

Soap bubbles, is a good first answer (◘ Fig. 4.3). But what are these beautiful bubbles actually made of. We already know neutral fats—composite lipid molecules that are electrically balanced, because the involved fatty acids—otherwise charged—are esterified with some sort of alcohol. This ester trick makes the negatively charged carboxyl groups of the fatty acids disappear. Maybe only transiently, because an ester bond can also be cleaved again relatively easily. Just take some aromatic olive oil, and boil it with some caustic soda, some sodium hydroxide—very quickly you get soap. Yes, "soap", which also is the chemical name for the salts of free fatty acids. These soaps have two completely different faces—a water-hating (hydro-*phobic*) and a water-loving (hydro-*philic*) part; and such dual behavior is pretty common among many fatty components.

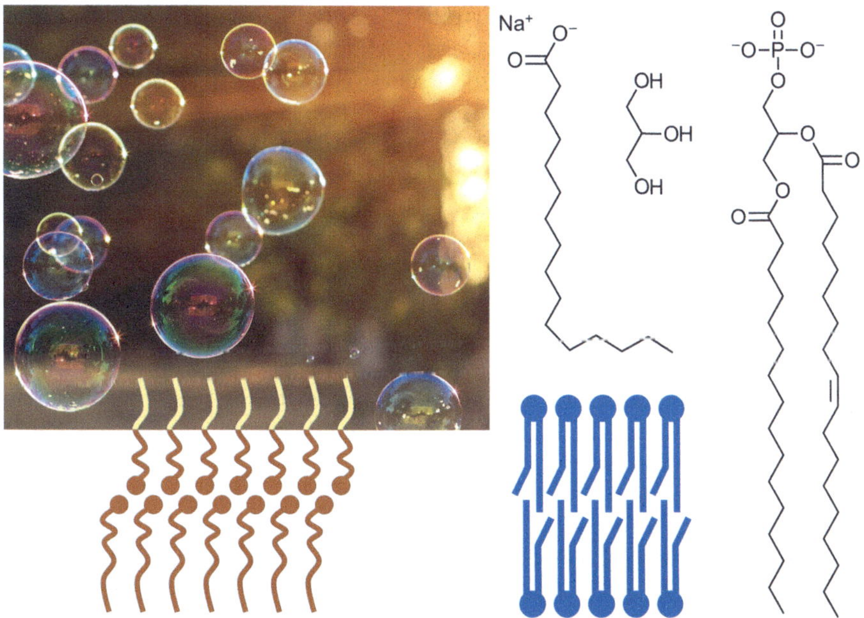

Fig. 4.3 **Soap bubbles and double membranes**. Those soap bubbles look beautiful in the evening sun. Common soaps are the sodium salts of fatty acids, such as palmitate shown here, the soap of the C16 fatty acid palmitic acid. In the thin outer film of a soap bubble, individual soap molecules form a rather thin double layer, with bulk water and the charged head groups hidden inside—the hydrophobic side chains however point outward to the air. Food for thought are membranes made of protein—be it stiffly beaten egg white or that foam that one can see at times at the sea. The lipids of biological membranes are a bit more complex. A widely spread version are phospho-lipids based on glycerol, with fatty acids and a phospho-head on it. Such lipids form double membranes in an aqueous environment, here now the polar head groups point outward to the watery environment. The lipid shown comprises a C16 palmitic acid residue and a C18 oleic acid residue. And this very oleic acid chain contains a double bond between its ninth and tenth C-atom. (Photo credit: Soap bubbles, © Lumppini ► stock.adobe.com)

Soaps can form soap bubbles. Within the outer film of the bubbles, let's call this film membrane for now, the water-soluble parts hide inside the membrane, the hydrophobic parts stick towards the airy outside. These double membranes look beautiful. For fans: Soap bubbles appear rainbow-colored, because their membranes are a few hundred nanometers thick, which is the same range of wavelengths of light, causing coherent light phenomena. Sometimes soap bubbles are also surprisingly stable, well, for many moments at least. Put into aqueous environments, the same building blocks that form soap bubbles, also might aggregate to form double membranes. Then, obviously, the membranes are the other way around—the hydrophobic tails point inwards and the hydrophilic heads present themselves to the outside world (Fig. 4.3).

How do **membranes** find their shape actually? Good question. First of all, it's the two-parted, amphipathic lipids themselves, that notoriously try to hide that one part that doesn't fit with the environment at that very moment. The driving force largely is the hydrophobic effect, again. Many factors influence how flexible a membrane is—be there long fatty acid sidechains or not, be there double bonds or not, is there a phospho-head group or something else. Think of dedicated flipase enzymes that unmix the lipid content in both halves of the double membrane—a double membrane with different lipid composition surely is is close to a piece of art. A membrane like this very quickly could bulge and spontaneously form more complex shapes.

If a membrane gets too wobbly, it might get close to rupture. The membrane should probably have been made from longer lipids—yes, but that wouldn't really be a short-term solution. The remedy is **cholesterol**. This magical bullet of membrane biology thickens membranes at higher temperatures; it stabilizes them. But cholesterol can do even more. At low temperatures, it also keeps membranes in their liquid state for longer. Without the good old "membrane lubricant" cholesterol, membranes would literally crystallize and then—very stiff—tear or break.

There are a great many proteins around to keep our membranes in check. Fun fact: A biological membrane may contain protein, in up to 50% of its weight. That is quite a lot of protein. There are flipases, for a start, enzymes that flip a component of the membrane from one side to the other, isn't that flipping exciting. There are stunning protein apparatuses for precisely controlled fusion of membrane vesicles; of cause other proteins also control the severing of membrane-wrapped deliveries. And there are a lot of transporters and translocases that pump all sorts of things between compartments, back and forth. Last but certainly not least, there is "the nuclear pore", a colossus of a protein complex with a high-tech polymer inside. Fun fact: With about 120 mega-Da the nuclear pore has the molecular weight of 30 80S ribosomes, thirty. Regrettably, I have to refer you to further literature, to learn more about all these exciting molecular machines.

If all living things were simply wrapped in soap bubbles derived from fats, why wouldn't all those droplets merge into a single huge lump of fat after a while? Especially during these times, where obesity is widespread on our planet (not referring to anyone specific out of the many authors of this book, well, me). Our cells prevent such a mess with the **cytoskeleton**. This is a meshwork made of fiber-like proteins or protein parts that provide mechanical strength. Some of these cytoskeletal proteins are involved in adhesion complexes, some sticky structures that allow a cell to attach itself to the extracellular matrix or other substrate. **Actin** is such a scaffolding protein. Strangely, the cell can build and re-build actin scaffolds, even under load. We prefer not to do so with our many scaffolds on normal construction sites, in the macroscopic world.

ℹ How stable are biological double membranes? And how great is the force that holds a protein in the membrane? While writing this book, I noticed that "we" life scientists are remarkably silent at this point. Please write to me if you know more about this topic or want to know more.

Compartmentalized Biological membranes are first-class self-assembly systems. They are stabilized by the hydrophobic effect. Lipid composition and packing effects between hydrophobic lipid tails determine the fluidity—or "wobbliness"—of membranes. Different membranes bend themselves to varying degrees. Cholesterol and many, many proteins significantly modulate the properties of biomembranes. Cells in our body, and eukaryotic cells in general, have a lot of membrane-enclosed reaction spaces; "Divide and Conquer" seems to be a winning strategy here.

- **Navigation**

We discussed the hydrophobic effect as a special biochemical driving force in ► Sect. 1.3.

We already know neutral fats from ► Sect. 2.3. Isoprenoids and cholesterol were mentioned in ► Sect. 2.5.

Pairs of the ATPase enzyme also bend membranes, as discussed in Sect. 5.7.

A comprehensive review article on membrane shapes. The authors believe that membranes are far more shaped by different lipid compositions than previously assumed—Bozelli & Epand. **2020**. Membrane Shape and the Regulation of Biological Processes, *J Mol Biol*. 432(18): 5124–36. ► https://doi.org/10.1016/j.jmb.2020.03.028.

4.3 Protein Folding and Why Small Molecules Make a Huge Difference

We can chemically link individual amino acids into peptides through peptide bonds. The ribosome can perform this fiddly work far better than us; the ribosome truly is a peptide factory. A long peptide chain, a polypeptide, leaves the ribosome… And then something astonishing happens: In less than a blink of an eye, this chain of amino acids takes on the shape of a functioning enzyme, of an almost mature antibody, or of some other sort of useful and much needed protein. Over many years, several brains have surrendered over the question of how protein folding would work.

It's all about the sheer number of possibilities in protein folding and the question: Does the polypeptide try out every possible conformation until one fits and then locks in? Even for relatively short peptide chains, the number of possibilities is huge; and it increases astronomically for longer peptides. **Napkin X** shows a simple wooden model of a protein together with approximations of theoretically possible conformations.

Napkin X: Studying Protein Folding at a Wooden Block Model

Here is a small wooden puzzle that I have been using for many years now in my teaching. 27 small cubes are lined up on a string. There is only one way to make a large cube out of it.

Between the 27 cubes there are 26 connections, **26 "bonds"**

Since everything in this model is square, each connection can assume 1 out of **4 states**

That's then 4^{24} to the power of 26 or **4.5 × 10^{15} possibilities**

So, even if I were to try out each possibility for just a blink of an eye, for about 0.3 seconds, for all possibilities to test I would need about **43 million years!**

Already in 1969, Cyrus **Levinthal** realized that protein folding cannot work by trying out all possibilities. It's quite remarkable that his name has become well known, "immortal" in a way, only through this thought experiment—through a **paradox**.

(Photo information: Wooden model, JWM, 2017)

Never ever does a polypeptide find its biological structure by stubbornly trying out everything. But how does protein folding work, instead? Perhaps we should start talking about energies—energies of folded and unfolded states. "Normal" proteins, for example from a human cell are **not particularly stable**. Only a few kcal/mol stabilize some proteins at 37 °C body temperature; and some of these proteins might already unfold during conditions of 40 °C, when you are down with a feverish cold. Proteins have been evolutionarily trimmed for their biological function, proteins have not evolved to be as hard as concrete or stone. Here are various analogies, to try to capture the nature of proteins: Proteins are like glass—a solidified, amorphous melt. If that's too technical: Proteins are like cooked spaghetti, left overnight in a pot—a soft, but quite stable mass of a certain shape. And anyone who has ever brought a protein to crystallize knows how protein crystals behave when crushed—they break apart like jelly.

But how do proteins find their shape? When proteins fold locally, they slide down a slope in a virtual energy landscape. That landscape may be rugged, creating the danger of dead ends during folding; more or less this landscape is shaped like a funnel—the **folding funnel**. A protein can assume its native form by sliding down through this funnel. A major driving force of this downhill ride, among a few other energy contributions, is once again the **hydrophobic effect**. In a watery environment, a polypeptide would always try to hide its large, hydrophobic amino acid side chains, within its inside. At the same time, it wants to show its best polar and charged sides to the outside. Quite a few proteins, when made chemically or recombinantly, are "kind" enough to fold autonomously, even in most simple aqueous buffer solutions.

This strategy doesn't work with many other proteins, though. Those proteins seem to crave the good ingredients of **cellular environments**. Primarily, such an environment contains many other proteins for companionship, including folding specialists, the folding helper proteins. Then, there are small molecules that might destabilize proteins (as chaotropic reagents) or stabilize (as osmo-stabilizers). A surprising discovery of recent years certainly was that the nucleotide ATP itself stabilizes proteins at concentrations it is normally found in cells.

Then there are quite a few proteins that not only need their special ligands or cofactors to function properly; these specific small molecules would also help the proteins to assume their native structure. Depending on a ligand, such as a small molecule that is willing to engage, has become a functional principle for **nuclear receptors**. How does this work? For a start, quite some nuclear receptors are receptors for steroid hormones and other hydrophobic substances. In the cellular nucleus, nuclear receptors act as transcription factors. Without their proper ligand, however, nuclear receptors tend to be unfolded and, fluffy and wobbly as they are, bound to and locked by folding helper proteins, such as HSP90. When the appropriate ligand comes along, the receptor folds around the ligand, well, this actually holds true only for the ligand-binding domain. The properly folded receptor detaches from the folding helper protein, and migrates into the nucleus. There, activated nuclear receptors bind to hundreds of sites in the genome, finally influencing the activity of many genes. To fold or not to fold, forms the basis for an important molecular switch here.

Have you ever looked at 3D representations of protein structures. I personally think they are beautiful. Experimentally determined protein structures can be found in the Protein Data Bank, the PDB. The name is misleading, as the PDB also contains many ligands, a few lipids, and more and more structures of nucleic acids, of RNAs and DNAs. PDB browsers are special viewing software to look at protein structures. People have individual preferences here, for one PDB browser or the other; pretty similar to the discussion of what the best internet web browser might be.

A playful gaming approach to protein folding might make students happy, you might think. There are the **Fold.it** games, which were started off by David Baker. Playing as a team, or in tournament mode, you learn to fold proteins and earn points as rewards; on your way through the game, you might also learn something about protein structure. Should you sign up to this, your creative energy benefits protein structure predictions—gaming for science. Folding prediction has massively changed since the introduction of AlphaFold—very likely you will have heard that term.

Close-up on 3D Protein Structures When looking at proteins or at least when playing with protein structures, you quickly come across three major structural areas in proteins, called secondary structures. There are some curly structures shaped like a corkscrew, others look like bent and stacked arrows (▣ **Fig. 4.4**). We will discuss the third structural category shortly. Corkscrews and arrows are schematic representations for the two most common structural elements of peptides. A peptide can wind up and form an **α-helix**. Alternatively, several peptide strands, stretched out, can stick to each other and form a **β-sheet**. If the ever-changing sequence of amino acids

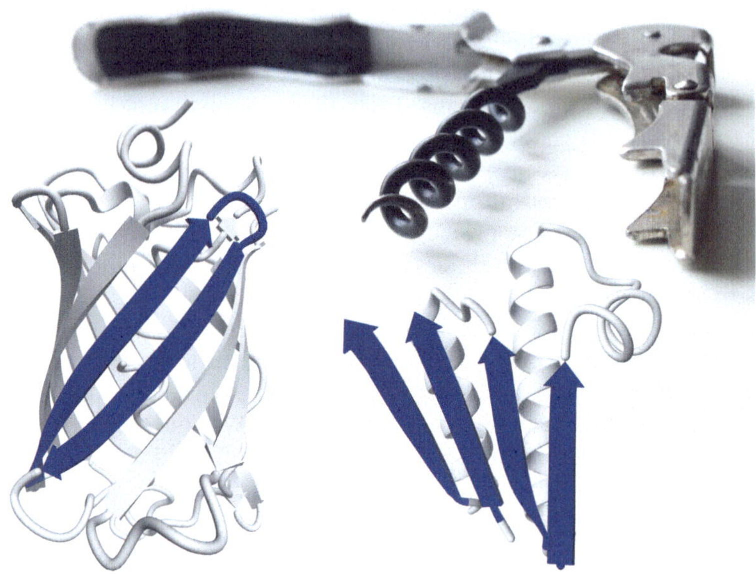

Fig. 4.4 **The secondary structure in proteins features corkscrews and arrows**. Protein 3D structures consist largely of twisted sections that resemble corkscrews (atop)—the α-helices. Other parts of the protein lie flat next to each other—stretched-out β-strands socialize to become β-sheets. β-sheets can either run in the same direction, forming a parallel sheet (bottom left), or in opposite directions, an anti-parallel sheet (bottom right). (Image credit: Corkscrew, JWM, 2020; Sheets, part of structures 1XNJ (left) and 1GFL (right); Visualization, JWM, 2022)

represents the primary structure, then we might call these tangles the secondary structure of a protein. But didn't we just mention three structural elements, not just two? There are also a lot of **loops**, cringles, and squiggles. They are not disordered in a direct sense, but they usually are highly individual for a certain protein; therefore, predicting these bespoke structural elements was incredibly difficult in the past. There are more layers of protein structure: the formation of domains might be regarded as tertiary structure, oligomerization of proteins the quaternary structure, and quinary structures are understood as somewhat looser interactions between full-blown proteins. Here, however, we will not go into detail about these three here.

Corkscrews Proteins and enzymes are something like cooked spaghetti made of long peptide chains, they are moderately stable, but they can be true champions of catalysis. A central driving force for protein folding is the hydrophobic effect. Parts of a polypeptide quickly engage in local interactions, allowing the protein to swiftly slide into energetic minima. Folded properly, water-soluble proteins usually have a hydrophobic core and a polar and charged surface. As fascinating as polypeptides can be, many proteins would be nothing without their cofactors, ligands, and other additions.

- **Navigation**

A lot has been written about AlphaFold. Please learn about this from other sources.

Amino acids and peptide bonds were covered in ▶ Sect. 2.2.

We will discuss champions of catalysis and other proteins in ▶ Sect. 5.1. More about the ribosome can be found in ▶ Sect. 5.2. Cofactors and their intimate relationship with proteins are discussed in more detail in ▶ Sect. 5.3.

ATP itself is not just a nucleotide, but also stabilizes proteins—Patel et al. **2017**. ATP as a biological hydrotrope. *Science.* 356(6339): 753–756. ▶ https://doi.org/10.1126/science.aaf6846.

More about kinases and their intimate relationship with their favorite substrate ATP—Brylski et al. **2021**. Cellular ATP Levels Determine the Stability of a Nucleotide Kinase. *Front Mol Biosci.* 8:790304. ▶ https://doi.org/10.3389/fmolb.2021.790304.

Confusing, a protein and its cofactor apparently enter the mitochondrial power plant via separate import pathways—Klein & Schwarz. **2012**. Cofactor-dependent maturation of mammalian sulfite oxidase links two mitochondrial import pathways, *J Cell Sci.* 125(Pt 20): 4876–85. ▶ https://doi.org/10.1242/jcs.110114.

Folding proteins can be fun, even in a seminar group. Learn more about the program Fold.it and the gamification of protein folding at Fold.it—Achterman RR. **2019**. Minds at Play: Using an Online Protein Folding Game, FoldIt, To Support Student Learning about Protein Folding, Structure, and the Scientific Process. *J Microbiol Biol Educ.* 20(3):20.3.63. ▶ https://doi.org/10.1128/jmbe.v20i3.1797.

4.4 Pre-sorting of Molecules and Pools Full of Metabolites

Shards of recycled glass and freshly harvested rice have something in common—both are mechanically sorted piece by piece, a very laborious process. In the lab, we do something similar when we do FACS. Times have long gone in which this meant to scan and send a certain piece of paper, then known as sending a *fax*. In the lab, FACS stands for "fluorescence assisted cell sorting", used to sort cell suspension for cellular size, granularity, and presence or absence of all sorts of markers—basically anything that could be detected with a specific binder, bound to a fluorescent dye. Whenever you want to sort piece-by-piece sounds laborious initially. But it doesn't have to be that hard, with a smart setup.

Within the cell, pre-sorting of metabolites is rather standard. The NAD and NADPH twins of cofactors are the prime example of separate pools. NAD and NADPH have a nicotinic acid moiety at their business end, and the special chemistry of this extension allows these cofactors to carry pairs of electrons. This is done massively in many biochemical redox reactions, within the cell. NAD and NADPH nearly look identical, these unequal twins however differ only by an innocent, small phosphate group, attached to NADPH far, far away from the nicotinic acid amide, the actual electron carrier. So what?

In fact, this tiny phosphate tag makes a huge difference within the cell (◘ Fig. 4.5), because proteins that are active in the electron transfer business can

4

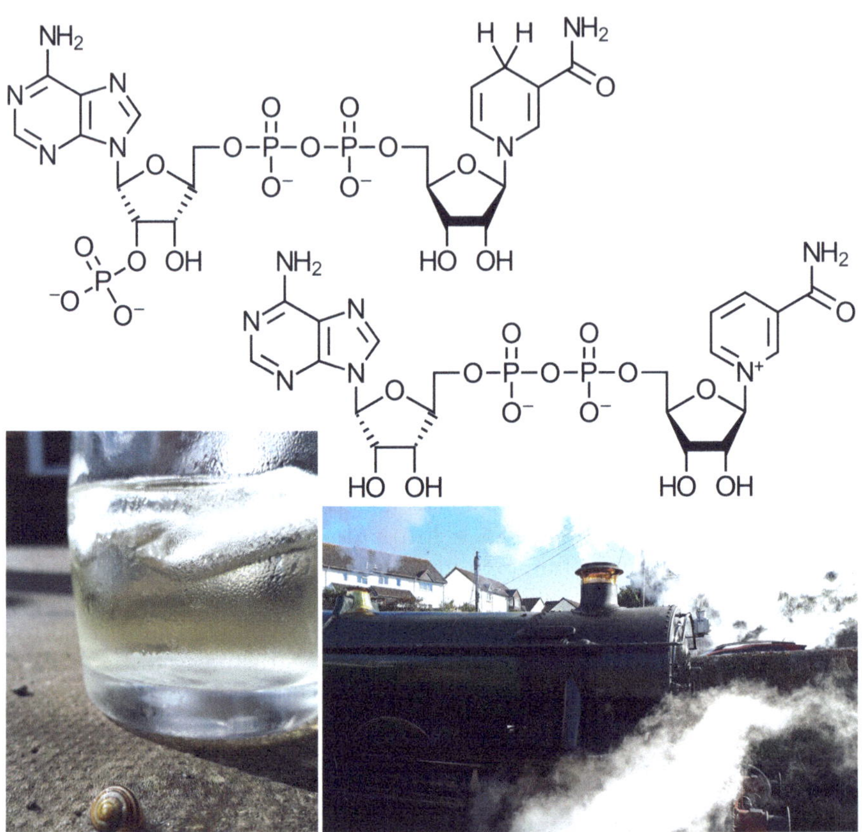

Fig. 4.5 **NAD and NADPH are like boiling water and ice cubes, in the same cup, at the same time.** The nicotinic acid amide allows NAD and NADPH to absorb electrons and to "store them temporarily", before releasing them again. Chemically, the cofactor twins work in exactly the same way. Biochemically, the phosphate residue at the 2'-OH of the ribose of NADPH—far, far away from the active part of the cofactor—makes a huge difference. The cell can distinguish the two cofactors very precisely—just as if you had hot coffee and ice-cooled gin & tonic in one vessel but could drink both separately of that same glass. (Image credit: Refreshment drink with ice, JWM, 2017, Steam locomotive in Minehead, Somerset, UK, JWM, 2018)

distinguish very precisely between NAD and NADPH. NAD-accepting so-called dehydrogenases are rather oxidizing proteins; instead, those dehydrogenases that run on NADPH are rather reducing enzymes. All these dehydrogenase proteins are redox catalysts, regardless of whether they oxidize or reduce. Biochemically important is that an average, metabolically fit cell constantly keeps NAD in a high surplus over NADH and NADPH over NADP. Technically, redox reactions are equilibrium reactions—whatever they involve. Coupling your special reaction to either NAD or NADPH, though, makes it practically a one-way reaction. This is hugely important to keep biosynthetic and bioenergetic pathways separate. The cell maintains large amounts of chemically very similar, but biochemically completely different electron acceptors and donors at very high AND very low energy levels; in the same compartment, at the same time. If you allow me to compare different redox potentials of electrons to different temperatures of water, this would be as if you could store boiling water and ice cubes, in the same cup, at the same time. Quite clever, such a cell.

Coexistence NAD and NADPH are electron carriers. They have almost an identical chemical structure but differ in a phosphate group on the end of the nucleotide handle. With enzymes being able to only bind one, but not the other, the cell achieves almost complete separation of two pools of redox equivalents. Other metabolites are separated by various modifications, by sorting into different compartments, or by other mechanisms. All in all, there is much more structure in the cell than you might think, when looking at its hodgepodge of many ingredients.

■ Navigation

In the cofactors in ► Sect. 5.3 we discuss NAD and NADPH from the perspective of the actual electron transfer.

How to Kick-Start a Chemical Reaction, and What to Do, If it Doesn't Work…

Contents

© The Author(s), under exclusive license to Springer-Verlag GmbH, DE, part of Springer Nature 2026
J. W. Mueller, *Ultimately Understanding Biochemistry*, https://doi.org/10.1007/978-3-662-71889-6_5

Both the kinetics of enzymes and the thermodynamics of biochemical reactions will be discussed in this chapter. Imagine: Saturday night, during early summer. The BBQ grill is stuffed with charcoal, the meat and the veggie burgers are ready to be grilled. …it just is, that the fire doesn't want to start. Something seems to be wrong with it. From a scientific viewpoint, this is not understandable at all. There is a tremendous amount of energy stored in charcoal. Thermodynamically speaking, the charcoal really "wants" to burn, but nothing happens. The coal definitely "wants" to undergo the chemical reaction with the oxygen in the air. But there is help. Firelighters, paper, wooden kindling, some cardboard as a fan, a hair dryer, or even a gas burner… any of these will likely get the reaction going, somehow—thus accelerating the reaction's course or its kinetics, while hopefully none of these will result in serious burns.

The above paragraph contains two major theoretical concepts. **Thermodynamics** describes the energy difference between the starting and ending point. It thus indicates whether energy can be released in a reaction at all—is a reaction energetically "worthwhile" or not. **Kinetics** on the other hand, indicates how fast a reaction takes place. Kinetics strongly depends on how high the mountain of activation energy is that needs to be overcome. It is kinetics that make a pile of wood chips smolder moderately over days, while producing delicious smoke flavor for curing ham or saussages. But it is also kinetics that make a handful of flour or sawdust explode, when mixed with just the right amount of air!

Enzymes, like any other catalysts, cannot do anything about thermodynamics, at least at first glance. But they can influence the course of a reaction significantly. In some clever way, enzymes might instantly establish the chemical equilibrium, and thus allow energetic coupling with other reactions. Alternatively, enzymes might open up a different reaction path towards the products; thus drilling a new path through the mountain of activation energy. What exactly happens when enzymes catalyze a chemical reaction is illustrated in ◘ Fig. 5.1.

◘ **Fig. 5.1** **What happens during catalysis?** Two scientists are discussing enzymatic catalysis. One has written a lot of formulaic stuff on the board. The other one points at the word *magic*, asking: "Could you possibly be a bit more explicit at this little detail?" I came across this joke at the 2007 FEBS conference in Vienna. Describing what's exactly happening during catalysis can be all but easy, sometimes one might best describe it as a miracle. Of course, that's not what is happening with enzymes at work. (Image credit: JWM 2022)

5.1 Enzymatic Catalysis: Hexokinase and TIM as Prime Examples of Proper Enzymes

An enzyme is a biological catalyst. An enzyme establishes the equilibrium of a reaction as quickly as it can. An enzyme will profoundly change the reaction kinetics; but it cannot influence the position of the equilibrium—it cannot scratch at the thermodynamic settings.

Our showcase enzyme is **hexokinase**. While consuming ATP, hexokinase coordinates the phosphorylation of glucose; but as its name suggests, hexokinase would happily accept a few other hexose sugars, as well. Hexokinase offers acid-base catalysis, at a supreme level. First, the enzyme tightly binds a molecule ATP. In fact, it doesn't bind naked ATP, but a magnesium-ATP complex. With the incoming sugar, hexokinase winds down into the catalytic state (◘ Fig. 5.2). The protein's lid and base rotate towards each other. The handover, the transfer, of the phosphate group happens in the direct vicinity of the magnesium ion, in its coordination sphere; the metal ion thus acts like a well-oiled joint here. Magnesium ions probably are involved in many, if not all phosphorylation events. Ta-dah, the phosphorylated sugar is ready in no time at all. Products, all change, while the enzyme opens up again, ready for another catalytic cycle.

The enzyme **triose-phosphate-isomerase**, affectionately also called **TIM**, is even more of a showstopper. TIM is a β-barrel, a barrel made of β-strands, decorated on the outside with α-helices (◘ Fig. 5.3). Due to its pretty look, similar-looking proteins have generally been named TIM barrels. TIM doesn't do that much—it just aligns its substrates and, for catalysis, employs a minor trick that we'll discuss

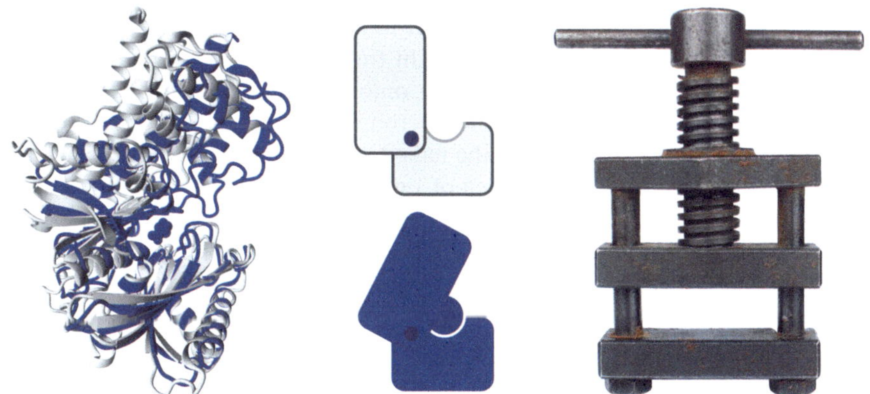

◘ **Fig. 5.2** **Hexokinase works like a thread press**. Two very similar structures of the enzyme hexokinase have been overlaid, only taking the "lower" part into account. The structure with glucose in the middle (blue ball model, PDB:1BDG) is significantly more bowed than the open structure without bound substrate (gray, PDB:1IG8). This nicely shows the enzyme's closure around the substrate for catalysis. This process might be reminiscent of a rotary press. (Image credit: own structure visualization of a structure of hexokinase from yeast ► https://doi.org/10.2210/pdb1ig8/pdb and a hexokinase structure from the parasite Schistosoma mansoni ► https://doi.org/10.2210/pdb1bdg/pdb; Rotary press, © aquatarkus, ► stock.adobe.com)

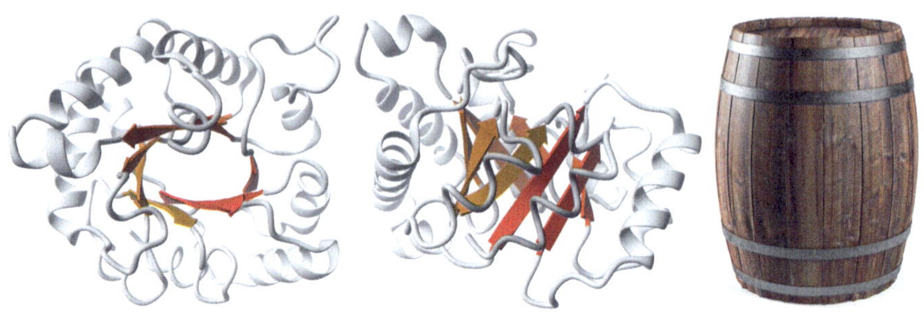

5

■ **Fig. 5.3** **A whole class of enzymes is named after TIM**. The enzyme TIM is remarkable. The enzyme triose-phosphate isomerase, called TIM, reaches an incredible speed during catalysis, due to a clever catalytic mechanism. All this machinery is plugged onto a notable protein fold, a barrel made up of beta-strands only, a beta-barrel. A lot of beta-barrel enzymes exist. This fold is also called TIM barrel. (Image credit: TIM, own 3D visualization of PDB ► https://doi.org/10.2210/pdb2YPI/pdb. Wooden barrel © simone_n ► stock.adobe.com)

later on. TIM converts an aldehyde into a ketone, and back again. For doing so, the enzyme stabilizes an en-diol state, the intermediate between a keto and an aldehyde form. Think of tossing a coin, TIM's mechanism could be comparable to stabilizing the state where the coin stands on its edge. TIM is considered a perfect enzyme, because the enzyme achieves breathtaking speeds. TIM is limited by nothing other than the diffusion of its substrates and products. One cannot catalyze any faster.

But how to speed-check an enzyme at all? An **enzymatic assay** is needed, that's an activity test, which transforms the substrate conversion into something measurable. Theoretically, such a conversion can be achieved by anything. Traditionally, however, reactions were preferred that involved a change of color. Think of frying onions in the pan, and how they change over time—they start off at some shade of white, and via translucent yellow, they might turn light brown, brown, and hopefully end up not too black. Indeed, frying onions is a chemical reaction, not an enzymatic one, but the point is that one can roughly follow the course of this reaction by change of color—visible to the naked eye. A photometer is a device that measures intensity of light, most often of only a narrow window of wavelengths; and these measurements tend to be pretty precise. Let's assume, we have a perfectly working enzymatic assay. If we now repeat that assay again and again with increasing amounts of substrate, while leaving the amount of enzyme constant, we see something unexpected: At low substrate concentrations, the rule of thumb "the more, the better" holds true. But the more the enzyme literally is bathed in substrate, the more saturated it is—"enough is enough" one might think. This behavior is further explained in ■ Fig. 5.4.

Whatever an ordinary enzyme does or does not do here, the **Michaelis-Menten equation** describes quite well that curve that we just measured by ourselves, well, virtually. It is pretty cool to have a not thaaat difficult equation, that describes what a lot of enzymes do. Let's look at this equation more closely. The speed v (*velocity*,

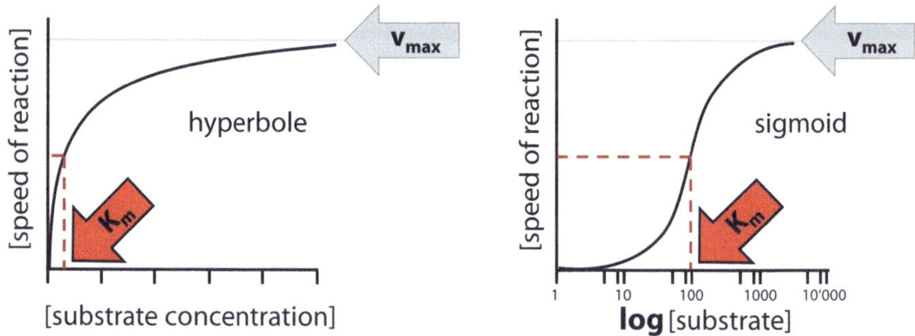

Fig. 5.4 **The course of an ordinary enzymatic reaction** When there is not much substrate around, every little bit more substrate leads to a higher speed of the reaction. But at some point, that's it. The enzyme is saturated. This behavior is best described by the curves displayed. A central point in these curves is the substrate concentration at which the enzyme has reached exactly half of its speed. This point is the half maximal effective concentration—EC_{50}. When describing the same phenomenon with the Michaelis-Menten equation, EC_{50} is also named the Michaelis constant K_M. The enzymatic behavior shown here can be described by a hyperbolic curve, as long as both axes are linear. When plotting the x-axis logarithmically, the hyperbolic converts into a sigmoidal curve. The two ways of presentation describe the very **same** reaction—the only thing that has changed slightly is the X-axis. (Image credit: JWM, 2022)

if you like), could simply be regarded as y. The maximum speed of an enzyme v_{max} is specific for the very enzyme we are looking at—it is quite high for TIM, as discussed above. What is left is the term "*[S] divided by the sum of* K_M *and [S]*". The little secret of this equation lies in this term—it represents a hyperbolic curve, that reaches half of its maximum at K_M.

$$v = \frac{v_{\max} \cdot [S]}{K_M + [S]}$$

The theory behind this equation was provided by Leonor Michaelis and Maut Menten (and we have a **meandering text box** dedicated to this special woman). They postulated that before an enzyme can turn over anything, it must first form a substrate-enzyme complex, maybe like this: $E + S \rightleftharpoons ES \rightarrow E + P$. Elegantly, they concluded that forming this complex—a biological binding event—is nothing more than a normal equilibrium reaction. And to such an equilibrium, the law of mass action can be applied.

If you want, you can deeply delve into the dark craftmanship of equation rearrangement for a moment, to derive the Michaelis-Menten equation from the law of mass action. And there are even two ways to get there, a thermodynamic one and a kinetic path, not sure which of the two is longer and stonier. Hopefully you will arrive at the Michaelis-Menten equation, in whatever way. Again, this formalism really is useful to describe the behavior of a lot of enzymes.

5

A Meandering on Mr. Michaelis, Mrs. Menten, and the Beginning of Enzyme Kinetics

Soon after Mrs. Maut Menten received her doctorate in medicine in Toronto, Canada in 1911, she decided to go to Berlin, Germany, and work with Leonor Michaelis on "enzymes", a very young and upcoming field of research. At that time, Dr. Michaelis was one of the world's leading experts on pH-buffered solutions. Dr. Menten certainly didn't have an easy journey over the Atlantic, as she set off across the ocean, only shortly after the sinking of the Titanic in 1912. Most probably, she travelled on her own expense.

Having arrived in Berlin, Maut Menten learned some German and worked in the hospital to make ends meet. Menten was a very energetic person, and it can be taken for certain that this power woman significantly pushed her colleague Michaelis, scientifically speaking. At this time, the doctor from Canada probably could not imagine that she will be mentioned in biochemical textbooks for her work on quantifying enzyme activity—perhaps forever.

Dr. Menten remained to be quite active in biochemical research, after her stay in Berlin. People considered Menten unstoppable and affectionately called her "dynamo". This is at least what people said in Pittsburg, her academic affiliation in later days. There was the saying that it was best not to cross her way, when she was driving. Maut Menten was promoted to full professor, only shortly before retirement.

This collection of articles about the life of Professor Menten certainly is worth a read—The Canadian Medical Hall of Fame. "Dr. Maud Menten". ▸ https://cdnmedhall.ca/laureates/maudmenten [Accessed on 3.6.2022].

The original paper on the Michaelis-Menten equation was published in German, an absolute classic of biochemistry, written by a Berliner and a Canadian. Particularly noteworthy is a recent translation of that paper into English, by the Dortmund-based Briton Roger Goody, who previously worked in Birmingham, England. Marvelous—Michaelis L, Menten ML, Johnson KA, Goody RS. **2011**. The original Michaelis constant: Translation of the **1913** Michaelis-Menten paper. *Biochemistry*. 50(39):8264–9. ▸ https://doi.org/10.1021/bi201284u.

The Michaelis-Menten equation actually only describes a molecular binding event—and many more things can bind to each other, than just an enzyme and its substrate. Therefore, a very similar formalism is used in pharmacology to describe the interaction between a receptor and a corresponding drug, be it an agonist or an antagonist. An advantage of the M-M equation therefore is its universality. Having said that, the description of "just" a binding equilibrium is also the weakness of this equation: The M-M equation might turn out to be insufficient to describe more complex substrate-enzyme complexes, maybe ones with added complexity of substrate inhibition or allosteric effects. For those cases, enzymology offers other models and other equations, such as the Hill equation for enzymatic twin enzymes (dimers, or multimers), that work cooperatively with each other.

Constant Now we know that enzymes are great. At times they surprise us with absurd catalysis rates, incredible precision, and 100% stereoselectivity. Enzymes become saturated at very high substrate concentrations. Well, one might argue that such concentrations may never occur in a real cell. Still, this behavior of enzymes can often be described very well with the Michaelis-Menten equation.

- **Navigation**

There is more about TIM in ▶ Sect. 5.5 to read.

Two Danes claim to have revolutionized enzyme kinetics, apparently, but then they describe linked enzyme-substrate complexes again as Michaelis-Menten-**like**—Dyla & Kjaergaard. **2020**. Intrinsically disordered linkers control tethered kinases via effective concentration. *PNAS*. 117(35): 21413–9. ▶ https://doi.org/10.1073/pnas.2006382117.

5.2 The Ribosome and Other Ribozymes

Okay, okay, proteins are great, they can do everything, and only they, they alone can do all that. That was probably the mindset of a lot of scientists, for a while—until Tom Cech described the first auto-catalytic RNAs in 1982—some sort of self-splicing introns. Maybe cool, but somehow exotic as well. It was the year 2000, in which evidence and opinion around enzymes made out of RNA, so-called ribozymes, changed dramatically. Based on a lot of preparatory work by Ada Yonath and other laboratories, Venki Ramakrishnan and Tom Steitz reported the successful structural elucidation of the ribosome, after decades of effort. By the way, the ribosome is THE site of protein production. It is a central component of every cell, whether that cell is from a bacterium, a primate, or a plant. The ribosome is huge, with more than two million Daltons. Ribosomal RNA and numerous ribosomal proteins make up the ribosome. And ribosomal proteins have beautiful names such as L3 or S6, depending on whether they are incorporated into the *large* or *small* ribosomal subunit. By quantity, nearly equal amounts of RNA and proteins make up a ribosome, maybe with a small surplus of RNA.

The ribosome is truly ancient. Absolutely every prokaryotic and eukaryotic cell contains ribosomes. And the inner bit of the ribosome—the binding sites and the peptide making center—is almost the same everywhere. This suggests that a primordial proto-ribosome must have been an obligatory component of LUCA. LUCA? Yes, LUCA, the *Last Universal Common Ancestor*. Luca is the primal ancestor of all life forms on Earth. Until recently, until today, parts of Luca are still alive in our cells; and the ribosome, without a doubt, is a prime example for this. When we look at the ribosome, we are thus looking billions of years into the past of life on Earth. The ribosome is a living fossil. Crocodiles are considered living fossils as well, as they have survived the dinosaurs and have hardly changed in the last two hundred million years. It seems that the ribosome and a crocodile have more in common than you might think (◘ Fig. 5.5).

5

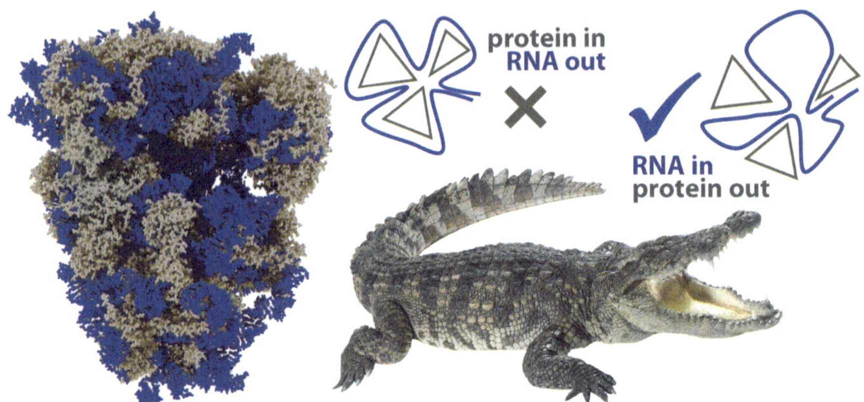

Fig. 5.5 **The ribosome is a crocodile**. You can see the human ribosome at 3.6 Å resolution here. The ribosomal RNA is shown in blue. All ribosomal proteins have a gray-green color. Occasionally, you might see small loose dots—these are supposed to be magnesium ions. The two models for the make-up of the ribosome are schematically represented, using the same color code. Until the year 2000, people argued whether the ribosomal RNA or the proteins are at the center of ribosomes. Many more 3D structures of ribosomes have been published since. They all confirm that the ribosomal RNA is in the catalytic center. The ribosome is a ribozyme—and a very old one indeed. Somehow comparable to a crocodile, which can be referred to as a living fossil as well. (Image credit: Ribosome, own structure visualization from ▸ https://doi.org/10.2210/pdb4UG0/pdb, JWM, 2022; Models, JWM, 2022; Crocodile © John Kasawa ▸ stock.adobe.com)

What exactly was reported in the year 2000 then? Tom Cech, yes, the one from those strange autocatalytic RNAs mentioned above, wrote a commentary, with the headline "The ribosome is a ribozyme". The interior of the ribosome, the catalytic center, where amino acids are cleverly connected to make peptide chains—new crystallographic studies showed that inner part at a resolution, never seen before. The catalytic center was clearly marked by a tightly bound inhibitor of protein synthesis. The real sensation was the absence of something—within 18 Å from the catalytic center there was no protein, no peptide, no amino acid, nothing. Absolutely nothing. The catalytic center of the ribosome consists of ribosomal RNA, exclusively (Fig. 5.5).

The ribosomal RNA itself catalyzes the peptide transfer. It aligns the substrates precisely to each other. Then a highly conserved adenosine (A2451) steals a proton from the amino group of the new amino acid to be incorporated. The previously completed peptide then is transferred on top of the newcomer. Obviously, this alone is already fascinating enough. In addition, though, various classes of known antibiotics target the ribosome. Hence, people hoped that development of ribosome-targeting antibiotics would receive a significant boost from these ground-breaking results.

Meandering: The Ribosome Is Good for an Oral Exam

A great press release, a smart acid-base catalytic mechanism, accomplished single-handedly by a few nucleobases. As a student, I was thrilled by the headlines. It happened that my main and final examination of my biochemistry degree course, was about to be due—an oral examination. I couldn't help but bring this topic into the exam. More precisely: I brought a printout of the paper with me. It also happened that the examiner had not yet heard of the story.... What can I say? The exam went well.

I can only recommend to every student ahead of advanced oral examinations, to bring with you a hot-of-the-press research paper into the exam that they personally find totally exciting... But please only do so, if the paper vaguely fits the topic of the exam... #GeekAlarm

Here the reference for fact-check: Tom Cech announces it, the sensational message—Cech. **2000**. The ribosome is a ribozyme. *Science*. 289(5481): 878–9. Commentary. ► https://doi.org/10.1126/science.289.5481.878.

Crocodile The ribosome is a genuine ribozyme. It is the ribosomal RNA that forms the new peptide bonds. Deciphering the structure of ribosomes was a huge task. The ancient ribosome provides insights into the very early development of life—an era of a putative RNA world. Structural differences between ribosomes of different organisms can now be exploited to develop new drugs.

▪ Navigation

Cool things can be done with ribosomes. In ► Sect. 2.6 we briefly mentioned ribosome display.

We have already talked about the RNA world in ► Sect. 3.7.

In ► Sect. 8.3 we will steer towards protein optimization. Here we go again.

The ribosome crystallographer Ada Yonath writes about the future of antibiotic development—Matzov et al. **2017**. A Bright Future for Antibiotics? *Annu Reviews Biochem.* **2017**. ► https://doi.org/10.1146/annurev-biochem-061516-044617.

5.3 Cofactors for a Special Chemistry

It is quite astonishing what proteins and ribozymes can achieve, with regards to catalysis. Nevertheless, the type of possible chemistry is somewhat limited. Isn't it always the spatial orientation of the reaction partners and some acid-base catalysis. These two probably cover most of the catalytic capabilities of the four nucleobases and even most of the twenty standard amino acids. Okay, a bit of redox chemistry happens at cysteine residues. But that's it then, really. Cofactors add spice to the biochemistry of the cell, enabling a rich and diverse set of metabolic reactions.

Some of these cofactors are metal ions, such as magnesium Mg^{2+} and zinc Zn^{2+}. Iron already comes across somewhat more elaborate, in various flavors of iron-sulfur clusters. The more complex cofactors, however, are of organic origin or at least contain a complex organic part with a metal ion in the center. A non-representative selection of biochemical cofactors is provided in ◘ Table 5.1.

A lot of cofactors consist of a **handle** and a **special part**—the business end—that is responsible for the actual biological function. Think of a combination tool system from the hardware store, maybe: A moderately standardized grip with the charging unit, and many different special parts, on the other side. Often, the handle consists of a nucleotide, sometimes two nucleotides are combined... Quite often the nucleotide "handle" comes from ATP (◘ Table 5.1). Because it is found nearly everywhere, adenosine, the adenosine handle, is referred to as a fossil from the very early days of biochemistry on our planet (◘ Fig. 5.6). Logically, adenosine needs to have emerged in a time BEFORE the emergence of the ribosome. But that shall be enough about the handle for now.

5

Often the biochemical "magic" happens at the other hand, at the **special part**, the cofactor's active site or *business end*. A good number of these bespoke parts are derived from vitamins (◘ Table 5.1). Actually, all B-vitamins are used to assemble cofactors of some sort. Niacin, for example, also known as vitamin B3, is urgently needed to produce the cofactors NAD and NADPH.

A common motif of cofactors is that they are supposed to carry something small, slippery, or biochemically weirdy. Cofactor "pincers" are there to hold that tricky stuff, and to precisely release it again—sometimes even in activated form. **NAD** and **NADPH** are formally able to transport a **hydride ion**. A what? That would be a negatively charged hydrogen particle. Surely, that doesn't exist! Wait a moment, the twins NAD and NADPH mainly care about transporting a pair of electrons, with a loosely associated proton. Actually, NAD and NADPH are carriers and storage of pairs of electrons, the poor proton has to look after itself (◘ Fig. 5.7).

A different **mini battery** is the cofactor FAD, this time accepting single electrons, only. Coenzyme A, folate, vitamin B12 or S-adenosylmethionine are responsible for the **transport of carbon units** of various lengths... Thiamine pyrophosphate, on the other hand, provides a fairly reactive **aldehyde** function... and so on... and so forth. Please remember: The 20 standard amino acids or 4–5 standard nucleobases could not realize all this chemistry.

ⓘ Cofactors make biochemical life colorful. My favorite cofactor is PAPS, a molecule related to ATP, that transfers sulfate groups. Do you have a favorite cofactor as well? And why? Please let me know.

Adenosine is something very old, biochemically speaking, and so are proteins that bind adenosine. The way protein domains bind adenosine in general and NAD or NADPH in particular, are ancient and have remained almost unchanged for a long time. "Learning" to bind one of the molecules of this section can quite suddenly

Table 5.1 Some cofactors in biochemistry (alphabetically sorted)

Cofactor	Business end/special Part	Handle	Function	derived from
ATP	Phosphate/ Pyrophosphate	ADP/AMP	Phosphate-Transfer	ATP
B12	Cobalamin	AMP	Methyl group-Transfer	Vitamin B12
CDP-Diacylglycerol	Diacylglycerol	CDP	Phospholipid-Biosynthesis	CTP
CoA	Cysteine, Pantothenate	ADP + 3‘-Phosphate	Acetyl-Carrier	Vitamin B5
FAD	Riboflavin	ADP	One-Electron-Transfer	Vitamin B2
NAD	Niacin	ADP	Two-Electron-Transfer	Vitamin B3
NADPH	Niacin	ADP + 2‘-Phosphate	Two-Electron-Transfer	Vitamin B3
PAPS	Sulfate	ADP + 3‘-Phosphate	Sulfate-Group-Donor	ATP
Pyridoxal-phosphate	Pyridoxal	Phosphate	Amino and Carboxyl groups	Vitamin B6
SAM	Methionine	Adenosine	Methyl-Donor	ATP
UDP-Glucose	Glucose	UDP	Sugar-Transfer	UTP

Please look up the full names of all these substances in standard biochemistry textbooks. Here, adenosine always refers to the adenine nucleobase linked with a ribose sugar. ATP, UTP, and CTP also stand for the corresponding ribo-nucleotides. Deoxyribose components are only found in DNA, never in cofactors

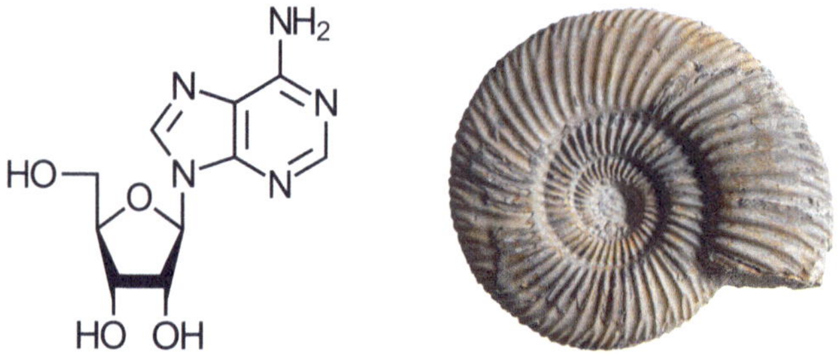

Fig. 5.6 Adenosine is a molecular fossil from prebiotic times. Adenosine seems to be everywhere—in DNA, in RNA, and in energy metabolism. Adenosine also features in quite a few enzymatic cofactors. (Image credit: © alice_photo ► stock.adobe.com)

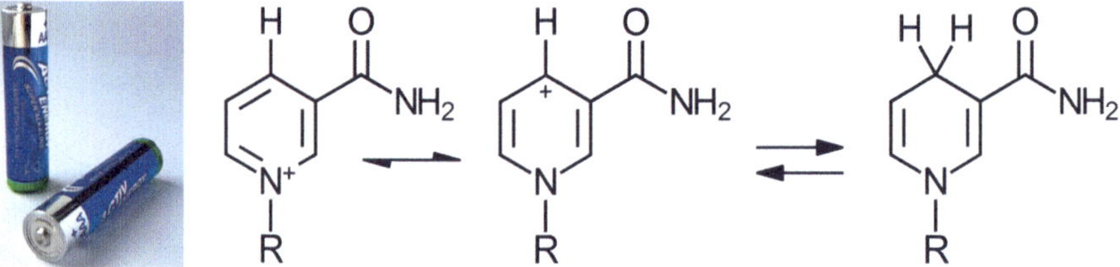

Fig. 5.7 How NAD and NADPH manage to transfer a "hydride ion". Chemically, the cofactor twins NAD and NADPH do their job exclusively with the nicotinamide head group. That nicotinamide has a nitrogen atom in this aromatic six-membered ring, and this feature makes this system of electrons somehow "poor", there simply isn't enough electron density for everyone in this ring. That nitrogen can balance its partial positive charge through mesomerism. But like a blanket that is too short, that compensation leaves the carbon opposite the nitrogen covered with too few electron-density. The carbamide at the side certainly helps, even if not directly involved in this process. What matters is that carbon opposite of the nitrogen partially converts to a carbonyl cation—something very, very rare in biochemistry. "Hungry" for electrons, this carbon gladly accepts a pair of electrons, well, and an accompanying proton. Remarkably, the ring no longer is aromatic, but at least there are enough electrons for all atoms of the ring for now. (Photo info: AAA-Batteries, JWM, 2022)

make a "boring" enzyme a world champion of catalysis. Enzymes just needed to get a suitable binding domain, for example by genetic reassembling of protein domains. Alternatively, enzymes just have to repurpose some nucleotide binding pocket, previous of some use otherwise. The result could be a novel dehydrogenase using NAD or NADPH, or a coenzyme A-dependent acetylase. Redesigning proteins in this way has happened again and again in the evolution of enzymes. We even might want to re-enact this process, during biotechnological enzyme evolution.

One of the most astounding cofactor-dependent stories of recent years is the one about the secret life of one of our histones, which now are supposed to be catalysts, in addition to being the hair roller proteins for our DNA. This story is about histone H3, for a long time believed to be a completely normal and well-behaved DNA-wrapping protein, a proper histone. But no, H3 has a secret other life as an NAD-dependent copper reductase. H3 keeps intracellular copper ions in check—they are kept at Cu(I) level; thus, avoiding copper in its toxic Cu(II) form. All of utmost importance for the intracellular redox balance. What A Scandal! Strange though that this story didn't made headlines in the tabloid press.

Cooperation Proteins and RNA are structurally diverse and capable of catalyzing quite a few astonishing biochemical reactions. At times, however, this catalysis is somewhat flat. Only cofactors spice up the biochemical soup of enzymatic catalysis. A lot of cofactors consist of a handle and some sort of special part, the business end. Nucleotides are often used for those handles. Evolutionarily, a protein can drastically change its function if it "learns" to bind a cofactor, or if it masters to switch binding to a different cofactor.

- **Navigation**

NAD and NADPH were already discussed in ▶ Sect. 4.4.

In ▶ Sect. 8.3 we will continue discussing protein optimization.

Adenosine as a biochemical fossil—Denessiouk et al. **2001**. Adenine recognition: a motif present in ATP-, CoA-, NAD-, NADP-, and FAD-dependent proteins. *Proteins*. 44 (3): 282–91. ▶ https://doi.org/10.1002/prot.1093.

Here is a more recent reference, documenting that adenosine is something VERY old…—Narunsky et al. **2020**. On the evolution of protein-adenine binding. *PNAS*. 117(9): 4701–4709. ▶ https://doi.org/10.1073/pnas.1911349117.

…and the News & Views article on the secret second life of histone H3 as an NAD-dependent copper reductase. Who would have guessed.—Rudolph & Luger. **2020**. The secret life of histones. *Science*. 369(6499): 33. Comment. ▶ https://doi.org/10.1126/science.abc8242.

5.4 The Dark Side of Steroid Hormone Biosynthesis, and Light at the End of the Tunnel—Vitamin D

In this book, we have already discussed a number of enzymes—for example, we showcased the hard-working screw press hexokinase and the good-looking TIM, the perfect enzyme. These enzymes are great biocatalysts. Primarily, they achieve catalytic mastery by perfect alignment of amino acid side chains and precise substrate binding. If you believe that all proteins are as great all the time, please join me diving into steroid biochemistry. From a biochemical perspective, there are at least three noteworthy enzymes around steroids:

1. There are the "neat" hydroxysteroid dehydrogenases, also known as HSDs. These enzymatic magicians can do what a lot of other enzymes cannot do. They selectively can catalyze either the forward reaction, or the reverse reaction, exclusively.
2. Then we have the "dirty" cytochrome P450 enzymes, also known as Cyp's, or just cyps. These enzymes' hard work is far away from being 100% specific. Oxygen is flung around wildly. Electrons are lost, here or there, all the time. A dirty and sooty business.

3. Steroids also lead us to the sunny side of life. A certain reaction during the making of vitamin D requires a lightning to strike, well, or at least some photon to cleave a special bond.

After all, steroids are a great thing. We already know that the basic steroid backbone is firm and very stable. A material, ideally suited to "engrave" molecular information into it. That molecular "engraving" mainly occurs by removing some carbons and/or oxidizing some other carbon building blocks. All to produce steroid hormones, powerful messengers which regulate many things in our body (◘ Fig. 5.8). Steroid hormones regulate our well-being as mineralocorticoids. The steroid hormone cortisol saves our lives in critical situations, it is our stress hormone. And there are the sexy steroids: Men and women always have both androgens and estrogens in their bodies at all stages of life, albeit in different ratios. It is the right mix that matters here.

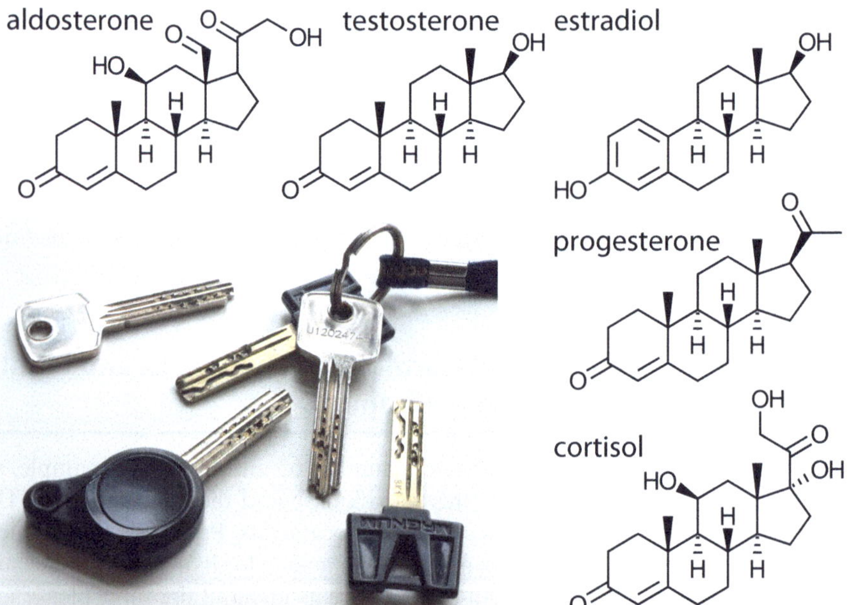

◘ **Fig. 5.8** **One can make all sorts of steroid hormone "keys" from cholesterol**. Powerful steroid hormones differ in their structure only at a few, but important sites. The structures (from top left) of aldosterone, testosterone, estradiol, progesterone, and cortisol are lined up here. The structure of the steroid-derived 1,25-vitamin D3 calciferol is shown in the next figure. Fun fact: If you line up all the names of the steroids in this picture, also including calciferol, and then order them alphabetically, their first letters together read A-C-C-E-P-T. Maybe suggesting to accept that steroids are important… By the way, the "forging" of steroid hormones is not for the faint-hearted… Especially the cyp enzymes often drop an electron. That's why the cortex of the adrenal gland is stuffed with many components of our anti-oxidation machinery. (Image credit: Safety key, JWM, 2020)

The cyp and HSD enzymes, the steroid key makers, share a common feature—they are not particularly specific. For "their" respective reaction, these promiscuous enzymes accept several similar substrates. When it comes to their actual reaction, these enzymes turn out not to be particularly precise. Thus, by-products are created every now and then. Some of these by-products merely were considered waste in the past. In recent years, however, biological functions have been discovered for some of these unusual steroids, this includes for example androgens oxidized at the C11-carbon, the 11-oxo-androgens.

The **hydroxysteroid dehydrogenases** (HSDs) can convert a hydroxyl group to a keto function or vice versa. They can precisely flip between these states, in a switch-like manner. Unlike most enzymes, hydroxysteroid dehydrogenases can specifically catalyze the forward reaction only, or exclusively the reverse one. HSDs can do so, because there are two distinct groups of HSD genes. Some HSDs are NADPH-dependent and catalyze the reduction of a certain functional group. Other HSDs are NAD-dependent, and these HSDs accomplish the reverse reaction. You might already know that a healthy cell almost always stocks a lot of NADPH and a lot of NAD in its cytoplasm, simultaneously. By selectively coupling to these different redox pools, these reactions become one-way reactions, always depending on which HSD enzyme is around at a given place and a given time, in a cell, or tissue.

Things get dirty with the **cytochrome P450 enzymes**, which we call cyps here. With their iron-heme group, cyps bind and activate oxygen—they are true oxygen catapults. Cyps do crazy things to steroids, they oxidize a side chain here, ablate a methyl group there, or aromatize a six-membered ring somewhere nearby. The whole making-of-steroid business is a pretty dirty affair, chemically speaking. During the process, a few electrons are always spilled and reactive oxygen species, ROS, are formed. The only other place where such dangerous oxygen radicals are usually formed, are our mitochondria, where cellular respiration occurs. Running steroidogenesis with all its cyps involved significantly increases internal ROS production. Any tissue or organ that wanted to engage with their own steroid production, therefore might particularly be exposed to high levels of reactive oxygen species.

Understandable that not every tissue in the human body is particularly keen to produce steroids for their own use, by themselves. A lot of tissues, however, would like to benefit from the growth-promoting signal transmission around steroids. A solution might be, to have a specialized body part for steroid production—some sort of a steroid factory? Yes, the liver is involved in this process, as always. With regards to steroids, however, the hotspot for hormone production is the adrenal cortex. The adrenal is referred to as a "redox-privileged" organ. In more practical terms, this means that basically everything that the human body has available to protect itself against free radicals is upregulated in the adrenal—for example, the production of glutathione and the enzyme superoxide dismutase SOD1. If we have agreed on the adrenals role as the steroid maker, how would we get the product then to the costumer? A little **meandering text box** might give the answer.

5

Meandering: How to Transport Steroids

A normal cell in our body would like to benefit from steroid signal transmission. However, that same average cell would prefer to avoid producing steroids on its own. Any mechanism by which steroids could be distributed throughout the body, would be most welcome here. The sulfation and desulfation cycle of steroids is such a **transport mechanism**, shown in the **Figure** in the **meandering**. Three types of enzymes are driving this cycle: (1) An enzyme that activates the stable, "lazy", sulfate ion. (2) An enzyme for the actual transfer of the sulfate group. (3) And one for sulfate cleavage in the periphery.

The free form of several steroids can only be found in very small amounts in the bloodstream, but their sulfated counterpart can be found there extensively. If you want, the sulfation of steroids could be seen as a temporary transport locking device. Sulfated in the adrenal cortex—"corked"—the steroid sulfate enters the bloodstream. Safely arrived at the end consumer cell or tissue, the "cork" is then removed, this can be a peripheral tissue like the colon or the brain. "Uncorked"—the steroid then can elicit its biological function, directly; alternatively, it can be further processed into a steroid hormone, further downstream the steroidogenic pathways. In all fairness, it must be added that steroids can be linked to other molecules as well, and that "free" steroids are not free. Actually, a lot of them are bound to certain carrier proteins such as albumin in the bloodstream.

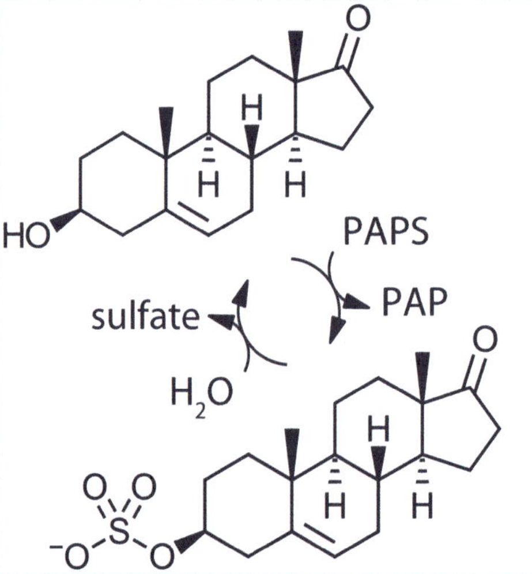

HO
H
H
H
O
PAPS
PAP
sulfate
H_2O
O
H
H
H
O O
S
$^-$O O

A thick review article with almost everything known on the topic of steroid sulfation, from the year 2015—Mueller et al. **2015**. The Regulation of Steroid Action by Sulfation and Desulfation. *Endocr Reviews*. 36(5): 526–63. Review. ► https://doi.org/10.1210/er.2015-1036.

(Image information: Open bottles, © Rafa Jodar ► stock.adobe.com; sealed bottles, © taveesaksri ► stock.adobe.com)

Maybe everyone knows that there is some connection between **vitamin D** and **sunlight**. But what is this connection about? And what is vitamin D anyway? Let's start with the last point. Vitamin D is a somewhat complicated biochemical entity. Actually, vitamin D is not a vitamin, since we can produce it ourselves; vitamin D is a hormone. Moreover, vitamin D is not a single substance, but a whole class of substances. Vitamin D3 originates in our own body, more about this in the next paragraph. In liver and kidney, it then is upgraded by a few hydroxylation steps, where again cytochrome-450-enzymes, cyps, are involved. This results in 1,25-hydroxyvitamin D3, a ligand that binds very tightly to the vitamin D receptor and thereby regulates numerous genes.

In vitamin D metabolism, one particular reaction is chemically exciting – it gets under the skin. Yes, within our skin vitamin D3 arises, when the weather is right for it. Only enough **sunlight** can cleave the B-ring of the pro-vitamin D3, also called 7-dehydrocholesterol. No enzyme can do this, because it is a pericyclic reaction, in which several pairs of electrons must migrate simultaneously. So far so good. This cleavage is tricky because the six involved electrons are not aligned with each other; they do not form an aromatic system; this is an anti-aromatic reaction. The migrating pairs of electrons are not compatible with each other. ◘ Figure 5.9 shows that two electrons are involved in the sigma-bond-to-be-broken; the p-orbitals of the

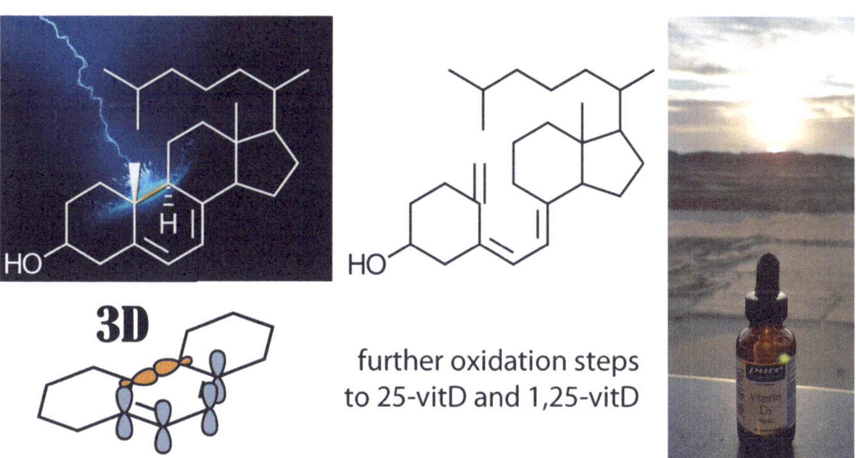

◘ **Fig. 5.9 Vitamin D—the sunshine hormone**. We start with provitamin D_3, also known as 7-dehydrocholesterol, highlighting the bond to be cleaved in orange, dramatically enhanced by lightning. This all is to make clear that cleaving "that" special bond of the vitamin D precursor, is a bit tricky, chemically speaking. A small cutout of the precursor is shown below, in a perspective representation, where the decisive electron clouds or orbitals can also be seen. This ring system is NOT aromatic; hence, its cleavage needs to be sun-kissed to occur. The end product of this light-dependent reaction is vitamin D_3—affectionally also called Cholecalciferol, depicted in the middle. I have drawn it in a way to better fit on the page. As added value, this representation clearly shows that Vitamin D_3 is derived from a steroid backbone. Further additions of hydroxyl groups later generate the actual active vitamin D ligand, involved in so many things. We close with a picture of some vitamin D in the sun. (Image credit: Lightning strike © Pavel ▶ stock.adobe.com; Vitamin D3 at sunset, Katwijk, NL, 2020; Reproduction with kind permission from Regine Niesen)

remaining four electrons are perpendicular to this bond. This reaction does not work enzymatically; heating it all up wouldn't work either. Only a suitable photon from the UV-B range of light can activate the bond between carbon atoms 7 and 8 and accomplish this reaction photochemically.

It's quite impressive that our body stores plenty of pro-vitamin D_3 in the outermost layers of our skin, waiting for a sunny day, only to retrieve the newly formed Vitamin D back from there. Within the biosphere, the photo-splitting of 7-dehydrocholesterol is more than a billion years old. Initially, the reaction probably served as sun protection only. Only when the vitamin-D-receptor evolved about 550 million years ago, did a hormonal system around vitamin D develop. Today, this system regulates immune functions and the calcium and phosphate balance in our body. Moderate sun exposure helps humans to meet their vitamin D needs, quite well. Hence, no need to recommend everybody a vacation in sun-kissed regions, for vitamin D synthesis only. If you live somewhere when the sun is low in the sky all day, or not there at all, problems might arise, however. Equally problematic can it be if one completely hides in their office all the time, or if one keeps their body fully covered at all times. At the very least, a balanced diet becomes important then, and dietary vitamin D supplementation might be advised.

Challenge If you think you've seen it all, biochemically speaking, looking at steroid metabolism might be refreshing for your mind. Most enzymes around steroids are not particularly selective, almost all accept multiple substrates. Some of the biocatalysts involved, the hydroxysteroid dehydrogenases, HSDs, then turn out to be one-way enzymes—they only catalyze the forward reaction, or the reverse, apparently not both. Other steroidogenic enzymes are far from being perfect. The cytochrome P450 enzymes involved here, cyps for short, are real soot throwers, they turn steroid production into a dirty business. There are glimmers of hope though. Among the steroids, an ancient light-driven reaction can be found. We humans can supply ourselves with vitamin D, by means of this photochemical reaction, if we are able to see the sun.

▪ Navigation

In ▶ Sect. 2.5, the waffle-baking enzyme lanosterol synthase was portraited during "steroid baking". The structure of cholesterol was part of ▶ Fig. 2.8.

More about the importance of protein promiscuity—Atkins. **2015**. Biological messiness vs. biological genius: Mechanistic aspects and roles of protein promiscuity. *J Steroid Biochem Mol Biol*. 151: 3–11. Review. ▶ https://doi.org/10.1016/j.jsbmb.2014.09.010.

Sulfation pathways hold much more interesting things than just transporting steroids, as this general overview might convey—Günal et al. **2019**. Sulfation pathways from red to green. *J Biol Chem*. 294(33): 12293. Review. ▶ https://doi.org/10.1074/jbc.REV119.007422.

A beautifully written review article about the vitamin D synthesis from an evolutionary perspective—Hanel & Carlberg. **2020**. Vitamin D and evolution: Pharmacologic implications. *Biochem Pharmacol*. 173: 113595. Review. ▶ https://doi.org/10.1016/j.bcp.2019.07.024.

More on vitamin D and other hormones in this educational review article—Lai & Mueller. **2025**. Understanding the biochemistry of hormones—message in a bottle. *Essays in Biochemistry*. 69(1):1–18. Review. ▸ https://doi.org/10.1042/EBC20240039.

5.5 The More, the Better—or—The Law of Mass Action, TIM, and Confined Spaces to React

Not a lot happens unsolicited in the world we live in. Most often a lot of persuading is needed. For example, when it comes to cleaning up an average child's room anywhere in Europe, or to de-clutter my desk, be it at university or in my office at home. Quite a few things have to happen for a cleaning-up actually to occur. A reaction with unfavorable thermodynamics, will never occur, no matter how fancy the kinetics might be—the best catalyst, which reduces the activation energy and thus speeds up kinetics, is of no use. Such "reactions" will never occur without a few tricks **to "optimize" thermodynamics**, more precisely we seemingly aim to trick with thermodynamics' second law.

A chemical equilibrium is like a small child: it tries to counteract the effect of the change, as far as possible. This is how Le Chatellier's principle describes chemical equilibria. The law of mass action captures the same statement, but in a more formalized way. "The more, the better" certainly is an approach to the topic that spontaneously resonates in many of us. At equilibrium, the forward and reverse reactions balance each other out. Intentionally changing the environmental conditions for this reaction, may be a control valve to push that equilibrium to this or that side. Chemists can adjust **pressure and temperature** for a reaction. Within the cell, there is not much you can do about pressure. But how about temperature? Just some food for thought: What exactly do we get fever for? (Spoiler: Maybe for our immune defense mechanisms to work a little faster…)

In biochemistry, water, or certain bits of it, often play an important role, during chemical reactions. The part of water involved might be a proton; sometimes it is a hydroxyl ion; and sometimes it's the H_2O molecule itself. **Water** or its components then fulfil two functions at the same time—water is the solvent for many reactions, it usually is available in more than sufficient quantity. But for the very reaction that we look at, water may also act as a metabolite; for parts of water, this holds true to a certain extent, as well. In biochemical teaching, it can therefore annoy some attentive students when a part of water spontaneously appears and then disappears again on the blackboard in the lecture, or even in some textbooks. The somewhat lame answer often is "we have enough of that" or "that now is included the constant". I hope I have not made that many "water errors" in my teaching.

For biochemical reactions, changing the **concentrations of metabolites** is an elegant screw for adjustment, possibly the adjustment parameter par excellence. Tapping on the concentration of a metabolite can completely reverse a reaction. That "pull" of a reaction, however, is only effective as long as the equilibrium is constantly re-established.

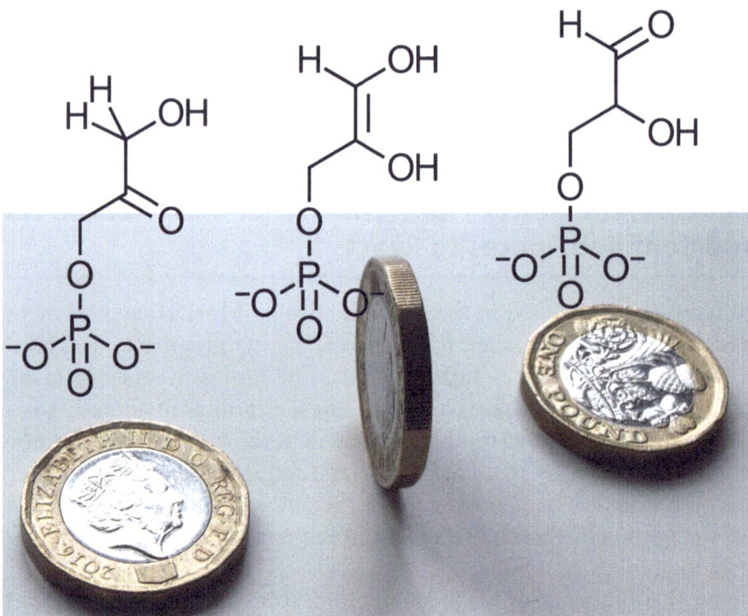

Fig. 5.10 TIM is pretty good at tossing coins. Triose-phosphate-isomerase, also known as TIM, brings di-hydroxy-acetone-phosphate (DHAP, left) and glyceraldehyde-3-phosphate (GAP, right) into equilibrium, and it does so in a flash. TIM achieves enormous rates of catalysis only by stabilizing an intermediate state, between its two substrates, an ene-diol. An ene-diol is a double bond that contains a hydroxyl group on either side. Think of flipping a coin, maybe. The ene-diol intermediate might be, as if the coin lands on its edge. From there, the compound randomly collapses down to GAP, or DHAP—randomly, but statistically obeying the equilibrium, falling down to DHEA more often than to GAP. Whatever, biochemically, GAP and DHAP are equivalent, thank you, TIM. (Figure information: Coins, JWM, 2020)

An example for two reactions coupled via shared metabolites comes from glycolysis: Imagine, aldolase just has gently split a doubly phosphorylated six-carbon sugar into two three-carbon sugars, the noble GAP (glyceraldehyde-3-phosphate) and the somewhat useless DHAP (di-hydroxy-acetone-phosphate). In downstream glycolysis, however, GAP is used, almost exclusively. So, if no clever biochemistry would be in place, half of the glucose you consumed, would go to waste. As this is no option within the cell, TIM is there to fix this—TIM is triose-phosphate isomerase (some people really abbreviate this to TPI, instead of TIM, don't ask…). What TIM actually does biochemically can be seen in Fig. 5.10.

So, TIM has to convert DHAP into GAP, understood. Like all glycolytic enzymes, TIM is strongly expressed in almost any cell. So, glucose comes down glycolysis, and TIM starts to establish the equilibrium between DHAP and GAP. It is a pity, however, that this equilibrium very much lies on the side of DHAP. The imbalance between DHAP and GAP, which is the actual chemical equilibrium, is at 20:1 for DHAP over GAP. Aldolase produces equal amounts of DHAP and GAP; but now TIM converts the "good" GAP into the "useless" DHAP due to the equilibrium. Wait, what's going on, is the enzyme working against us!?

Relax, as TIM's special qualities come into play now—**TIM is a perfect enzyme**, meaning that it converts DHAP and GAP into each other so rapidly that the speed of this enzyme is only limited by the diffusion of its substrates and products. TIM's catalytic constant k_{cat}/K_M is higher than $\mathbf{10^8\ s^{-1}\ M^{-1}}$. That's pretty fast—but how fast exactly? For an archaeal TIM enzyme, record values of 57.5 million turnovers-per-second for the conversion of GAP to DHAP were reported not too long ago! Of course, some might say: Wait, that TIM was a heat-loving enzyme, and that measurement maybe had been carried out at a temperature of 90 °C, or so. Just to put this into perspective: the CO_2-fixing enzyme RuBisCO from photosynthesis catalyzes the docking of carbon dioxide to ribulose-bis-phosphate at a speed of one to ten CO_2 molecules per catalytic center per second. Keeping in mind RuBisCO, it becomes clear that TIM really works at a breathtaking speed. Admittedly, the comparison is a bit messy, as TIM catalyzes a reaction on a single substrate, while RuBisCO has two substrates, and one of them, CO_2, even is gaseous. But still.

◘ Figure 5.10 explains how TIM could rise to be a world champion of catalysis, at least to some extent. TIM keeps the two three-carbon-sugars GAP and DHAP in equilibrium, forever and ever. This equilibrium works very much towards the principle of Le Chatellier. No matter how much or how little DHAP and GAP are present, they are always in a 20:1 ratio of DHEA over GAP. Whenever GAP is consumed, it is immediately replenished from the existing pool of DHAP. Should the cell run low in glucose, the subsequent reaction could gradually deplete any existing GAP and DHAP. This subsequent reaction is catalyzed by GAP dehydrogenase (GAP-DH) and is about converting GAP into 1,3-bis-phospho-glycerate (1,3-BPG), and some biochemically clever side business.

Why does this work? For tapping the GAP/DHAP equilibrium and to re-wire it, so that this reaction quantitatively proceeds in the direction of GAP formation, and nowhere else, some payment needs to be made, energetically. This energy comes from the subsequent reaction—the GAP-DH reaction, just mentioned above, is strongly exothermic. The aldehyde GAP is partially burned in this reaction; it is oxidized to the carboxylic acid 1,3-BPG—there really is some OOMPH behind it. The released energy is not lost. Instead, this energy is cleverly used to couple a phosphate ion to the "burning" of a part of GAP. Ta-dah, 1,3-bis-phospho-glycerate (1,3-BPG) is made. As a side product, this oxidation reaction also produces reduction equivalents in the form of NADH. Actually, this GAP-DH reaction is quite amazing, in many ways.

By **confining reactions** to certain compartments, coupled reactions can be increased in their efficiency. In the mitochondrion, reactions have to take place in a much more packed environment than in the cytoplasm—and this could even bring thermodynamic advantages. Some organisms are more special than others. Trypanosome parasites, for example, have a dedicated organelle, the glycosome, where almost all the reactions of glycolysis take place. The corresponding enzymes, including TIM, are provided with special address labels, so-called sorting signals, and sent to the glycosome. There, they have to perform their work altogether in a confined space.

Separating out confined reaction spaces, does not need membranes—larger protein complexes can parcel out individual reactions even within the same cellular compartment. The prime example for such confinement is channeling—when a "fragile" metabolic intermediate is passed on from one enzyme to the next one, without mixing with the bulk of the solvent around it. Such a kiss between two enzymes—a somewhat short-lived protein interaction—has long been suspected between the enzymes TIM and GAP-DH. Instead, however, an interaction between GAP-DH and phosphoglycerate kinase PGK, the next enzyme from within the glycolytic pathway, was experimentally demonstrated recently. The binding strength between GAP-DH and PGK—expressed as their dissociation constant K_D—was found to be in the range in which the proteins themselves were present in the cell. Somewhat technically, this means that you could make or break this interaction, gentle in nature, simply by slightly tweaking the volume of the compartment, be it minor changes to the water and salt content, making it swell or shrink; a more drastic volume change might happen during mitosis and cell division. Purely thermodynamically coupled reactions might respond to all of the above.

Coexistence A reaction that doesn't really want to proceed on its own, can be coupled to an exothermic reaction. There would not be life on Earth if such thermodynamic couplings did not exist. One type of coupling are shared metabolites that participate in subsequent equilibrium reactions. The glycolytic reactions GAP $\rightleftharpoons$ DHAP and GAP $\rightarrow$ 1,3-BPG, catalyzed by TIM and GAP-DH, respectively, are a prime example of thermodynamically coupled reactions. When the reaction volume is reduced at the same time, the efficiency of such a coupling can be increased further. For two substrates to react and the corresponding enzyme, the volume effect can be explained like this: It makes a massive difference for the three particles, if they are far apart, like for example three dancers in a spacious ballroom in Blackpool, or whether they are squeezed tightly together, like in a Birmingham nightclub.

Navigation

We have discussed water extensively in ▶ Sect. 1.1. But it's always good to refresh the peculiarities of water another time. The "jumping" protons, for example, make proton-dependent reactions extremely fast.

An overview of glycolysis and its highway-like character can be found in ▶ Sect. 6.2.

More about transaldolase, an enzyme related to aldolase, is presented in ▶ Sect. 6.4.

We will discuss the "lazy" enzyme RuBisCO in ▶ Sect. 7.3.

Why organisms have been getting fever for 600 million years—is explained here. Fever heats up the immune defense—Evans et al. **2015**. Fever and the thermal regulation of immunity: the immune system feels the heat. *Nat Reviews Immunol.* 15(6): 335–49. ▶ https://doi.org/10.1038/nri3843.

Faster, higher, further. TIM is fast, but other TIMs are faster. The fastest TIMs can be found here—Sharma & Guptasarma. **2015**. 'Super-perfect' enzymes: Structural stabilities and activities of recombinant TIM from *Pyrococcus furiosus* and *Thermococcus onnurineus* produced in *Escherichia coli*. *Biochem Biophys Res Commun*. 460(3):753. ▸ https://doi.org/10.1016/j.bbrc.2015.03.102.

Now it is measured, officially, the interaction of GAP-DH and PGK depends on the congestion- and mucus level of the cytoplasm—Sukenik et al. **2017**. Weak protein-protein interactions in live cells are quantified by cell-volume modulation. *PNAS*. 114(26): 6776. ▸ https://doi.org/10.1073/pnas.1700818114.

Ron Milo looks at metabolic pathways, thermodynamically, and discovers kinetic potholes—Noor et al. **2014**. Pathway thermodynamics highlights kinetic obstacles in central metabolism. *PLoS Comput Biol*. 10(2): e1003483. ▸ https://doi.org/10.1371/journal.pcbi.1003483.

5.6 ATP Is Molecular Dynamite—or—Coupling to High-Energy Molecules

One gram of dynamite (a mixture of 40% nitroglycerine and 60% diatomaceous earth) releases about 1.7 kJ of thermal energy upon detonation. When a single phosphate is split off from ATP, leaving ADP behind, 30.5 kJ/mol of heat is released. Converted into joules per gram (simply divided by the molar mass of ATP) this is 60 J/g or about 1/30 of the energy of dynamite. The high-energy molecule 1,3-BPG, discussed in the last chapter, already generates 232 J/g or 1/7 of dynamite. And for ATP and 1,3-BPG, we only consider the energies that are released when special bonds are split—unlike dynamite that completely goes up in heat and smoke.

Other molecules contain even more energy, our metabolism cannot release this energy as quickly, however. Through respiration, the body can release about 16 kJ of energy from one gram of glucose—this is almost 10 times more than the energy from one gram of dynamite. Fats even have a higher energy content. Biochemical mechanisms ensure that these processes run safely and smoothly. Nothing explodes, or hisses, or smokes, within our cells; well, maybe except for a few sooty cytochrome P450 enzymes.

A lot of reactions are driven by "burning" a **high-energy molecule**. And one of these high-energy molecules is 1,3-BPG, mentioned above, which is the product of the GAP-DH reaction (◘ Fig. 5.11). 1,3-BPG is an acid anhydride of a phosphate

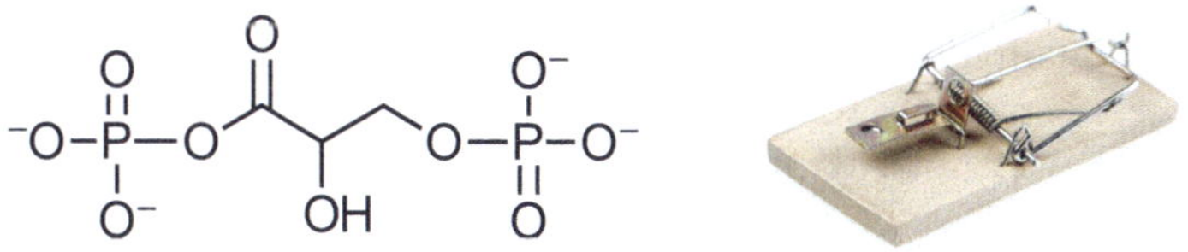

◘ **Fig. 5.11 The high-energy molecule 1,3-BPG and a mousetrap**. 1,3-bis-phosphoglycerate (1,3-BPG) actually is an acid anhydride from two different acids, the 3-phospho-glyceric acid on the one hand and phosphoric acid on the other. The 1,3-BPG molecule is stretched almost to its breaking point—within the cell, it carries four negative charges. If water severs the acid anhydride bond, at least one additional negative charge would occur, which makes the reverse reaction virtually impossible. (Image credit: Mousetrap © Stillfx ▸ stock.adobe.com)

group, on the one hand, and glycerate-3-phosphate, on the other hand. The 1,3-BPG molecule carries four negative charges at physiological pH—a small, but highly strained compound. When the anhydride bond is severed, the individual parts fly apart, immediately, due to electrical repulsion. Then, the split products engage in subsequent reactions: here, they deprotonate, there their electron shell resonance-stabilizes. All this makes the reverse reaction VERY unlikely to occur. I hope that the high-energy nature of this molecule becomes understandable, in this way.

Some biochemists like speaking of a "high-energy bond" here. Some textbooks even represent special bonds with a "~" tilde symbol. In this thinking, a bond is broken, and a lot of energy is released—possibly comparable to a triggered mousetrap. I think this concept is somewhat misleading. That spring in the mousetrap can only store all that energy because it was screwed onto a small wooden board and hooked into a trigger mechanism. Left alone in a vacuum, that very spring can't do a lot. The same applies to the bonds between the atoms phosphorus, oxygen and a second phosphorus atom in ATP or the phosphorus-oxygen-carbon bond in 1,3-BPG. Only the entire package of the pyrophosphate group or the whole 1,3-BPGs molecule makes the splitting of these bonds energetically so rewarding. In this book we always talk about whole high-energy compounds or molecules.

That other high-energy molecule, **ATP**, is considered the **universal energy currency** of the cell, by many. But what makes ATP well suited for this function? For a start, ATP is so great because it features an adenosine moiety. Admittedly, this **molecular handle** appears in many cofactors, so that we may conclude that it is easy for proteins to hold onto this adenosine "grip".

Secondly, compared to 1,3-BPG, ATP actually is not such a high-flying high-energy molecule—it is **medium-energy** enough, so that it can be produced using other high-energy molecules, such as 1,3-BPG. Good transport molecule better have "medium" affinity for the transported goods. Seemingly, a wider rule—thinking of hemoglobin, maybe, which should bind oxygen tightly, but only tight enough, so that the oxygen can be released again in target tissues.

Thirdly, ATP can be used in surprisingly many ways within our metabolism—most often, energy coupling with ATP works by transferring some part of the ATP to the molecule that needs an energetic kick. ATPs versatility reminded the author of a **Swiss pocketknife** (**see** ◘ Fig. 5.12). It is practical that different parts of ATP can be transferred: On the one hand, ATP has the chain of three phosphates, affectionally called α, β and γ. Briefly, we have mentioned the scenario where the γ-phosphate is transferred, which already contributes 31 kJ/mol of energy. If ATP was involved in the reaction catalyzed by TIM, that reaction's energy debt. would have been paid for easily here. ATP can provide more energy when the bond between the α and β phosphate is split and the AMP part (essentially the other side of the molecule) is transferred to the reaction partner. This reaction alone already provides more energy than the mere γ-phosphate cleavage. Even more energy comes from the subsequent splitting of the other reaction product—double phosphate (also called pyrophosphate). This type of reaction regu-

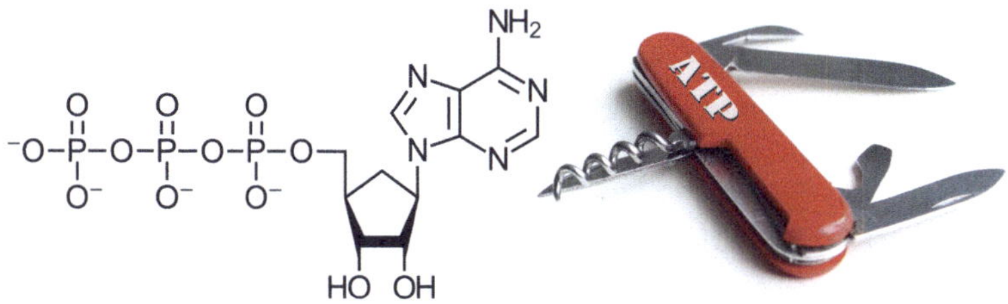

Fig. 5.12 **ATP is the Swiss pocketknife of metabolism.** ATP is **the** mediator between anabolism and catabolism. Usually, a single phosphate residue is transferred. Sometimes, however, a pyrophosphate is cleaved off, or an AMP residue is transferred, or even the adenosine nucleoside, alone. In this flexibility of use, ATP can certainly compete with a Swiss pocketknife. (Photo info: Our neighbor's pocketknife, JWM, 2020)

larly activates otherwise "lazy" metabolites, such as fatty acids, amino acids or the extremely inert sulfate ion.

Other reactions of ATP are known (e.g. ADP-ribosylation or AMPylation), they will not be discussed here, here. ATP is a truly versatile molecule. Universal, constantly present and always used. But how much is "always used"? On **napkin XI**, we will estimate how much ATP our body makes every day.

Napkin XI: The Turnover of ATP in the Body Is Really Remarkable

Our body doesn't contain a lot of ATP, those few grams of ATP however are constantly used up and regenerated again.

In our body, this happens at a rate of about **9 x 10^{20} molecules per second**

That's about 65 kilograms of ATP per day

All this is just at rest. During physical activity, ATP turnover can easily increase tenfold.

I took these numbers from a marvelous commentary by Peter Rich from 2003. There, he really pushed this calculation to the limit in all its marvellous detail—Rich. **2003**. Chemiosmotic coupling: The cost of living. *Nature*. 421(6923):583. ▶ https://doi.org/10.1038/421583a.

Concentrate We refer to high-energy molecules as compounds where there really is a bang, when a single bond is split. The energy of high-energy molecules is often used for group transfer reactions. Very often the transferred group is a phosphate, which often comes from ATP. Here, we discussed 1,3-bis-phospho-glycerate and ATP. There are other high-energy molecules and some of them really "mean it", like phosphoenolpyruvate, or PEP for short. "Burning" high-energy molecules in general, and specifically ATP, is often coupled to thermodynamically reluctant reactions—for that special kick.

Navigation

In ▶ Sect. 5.3 we have already mentioned that special adenosine grip, that is found in so many biochemical molecules.

5.7 Coupling to Gradients, the OxPhos Wars, and an Advanced Class in Molecular Cheating

5

By now, we know that energy-consuming reactions can be coupled directly to energy-yielding reactions. Most often, this coupling is mediated by ATP or another high-energy molecule. For many years of the 20th century, the search was on for **that** high-energy molecule that would drive ATP synthesis within the cell.

In 1961, Peter Mitchell postulated that the energy-consuming ATP synthesis could also be driven by a proton gradient across a proton-tight membrane. Mitchell caused quite a stir for many years, with these unorthodox views that seemingly offended some with more traditional believes. Mitchel was mocked, and even laughed at, for quite a while. Today we know that any tiny difference in concentration could represent an energetic potential that only would wait to be cleverly utilized. The energy hidden in whatever concentration difference, can be approximated using the Nernst equation (see **napkin XII**). Back to Peter Mitchell: in 1978 he was honored in Stockholm for his ideas, and some refer to this moment as the end of the biochemical wars over oxidative phosphorylation. History of science now knows these fights as the OxPhos Wars—not sooo long ago, in a galaxy not thaaat far away.

Whenever whatever biomolecule is distributed unevenly across any sort of diffusion barrier (such as a membrane), this concentration gradient represents an energetic potential. This biological battery "just" needs to be cleverly tapped into. A special transport protein for that specific biomolecule, connected to some sort of a biological dynamo, could probably be all you need. Distantly similar, an Ecofan is an amazing little device that works in a comparable way. It is explained on **napkin XII**.

Napkin XII: Can an Ecofan Help Reduce the Fear Against the Nernst Equation?

The Nernst equation distantly resembles the Henderson-Hasselbalch equation. The author has to admit a certain respect for the Nernst equation.

$$E = E_0 + \frac{RT}{zF} \ln \frac{[state\ one]}{[state\ two]}$$

But please look at this equation more closely: E_{zero} just is the potential of some standard—for fans: the standard of all standards is the standard hydrogen electrode. Any change of the state of the system-under-scrutiny then is measured relative to this standard. In practice, RT/zF is just another constant that is specific to the gradient we are currently looking at. More details about the Nernst equation can be found in dedicated textbooks. What remains is the ratio of the actual concentration gradient, expressed as the natural logarithm thereof. All now is nicely and obediently put together, for us to calculate the energy that could possibly be released if you tapped into that gradient. Ta-dah! Dealing with this equation turned out not to be that difficult in the end.

The Nernst equation shows that every concentration difference of a given biomolecule—absolutely everyone—has the potential to be biologically exploited for energy production. An Ecofan might be a somewhat useful analogy for this argument. An Ecofan is an amazing device—it consists of two aluminum parts that are thermally decoupled by a thick insulating layer. When the Ecofan is placed on a hot surface, such as a stove, the lower part quickly heats up, while the upper part more or less remains at room temperature, as it can dissipate heat effectively due to its design, resulting in a large surface area. The temperature difference between the two metal parts generates a voltage difference, large enough to power an electric fan. Many people who have a log burner in their house also own such an Ecofan.

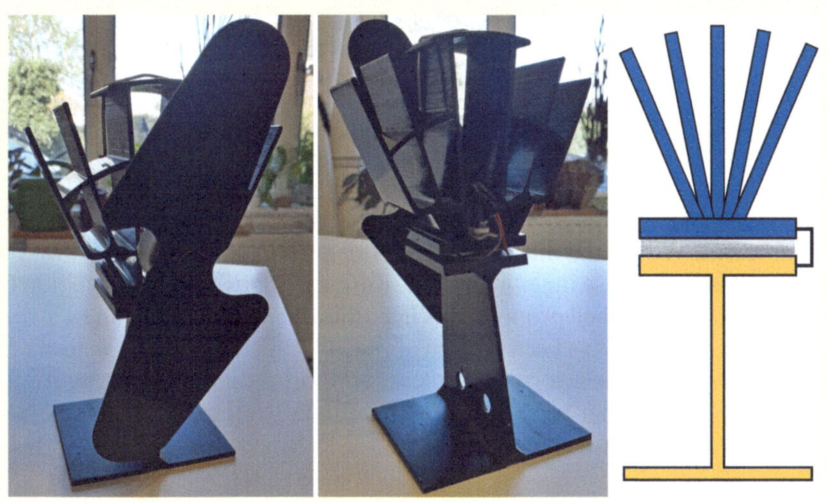

(Image credit: Ecofan, JWM, 2021)

The cell's Ecofan is the ATP synthase. Its membrane part sits in the proton-tight inner membrane of the mitochondrion—this part historically is named F_O. The other part, protruding from the F_O part, into the mitochondrial matrix, is known as the F_1 part. Voila, there you have the F_1F_O-ATPase. A beautiful example for "useful" nomenclature in biology: the F-"one" part got its name from a "fraction 1" of some biochemical purification, a long, long time ago. The F-O part, however, carries the letter "O"—as in Oliver—in its name (no, it is not a zero), because F_O's activity can be inhibited by the oligomycin antibiotics. Yeah, I cannot stress enough how important consistent and logical nomenclature can beee (hope you sense the irony here…).

Much easier to remember are ATPase's nicknames—the enzyme is also known as *Fifo*-ATPase; or even as lollipop-ATPase, lollipop because of its 3D structure, which—at low resolution—resembles a lollipop (Fig. 5.13). The *Fifo*-ATPase

5

Fig. 5.13 **Lollipop-ATPase**. Strange balls-on-a-stick must have been what researchers looked at in early electron microscopic images of mitochondria. Maybe legitimate reasons for ATP synthase to receive its nickname. Pictured here is the bovine *Fifo*-ATPase enzyme in two different renderings, the stator is not shown, however. (Image credit: own structure visualization after ▶ https://doi.org/10.2210/pdb5ARE/pdb; Lollipop, © Gelpi ▶ stock.adobe.com)

enzyme couples the energy-consuming synthesis of ATP to an energy-providing proton gradient. Within the mitochondrion, this proton gradient is stepwise built up across the proton-dense inner membrane during oxidative phosphorylation (see ▶ Sect. 6.3). When Mitchell proposed his hypothesis, *Fifo* was still a long way from being discovered. Much has been written about ATP synthases since. One characteristic of *Fifo* is that it is a sociable protein. *Fifo* forms dimers, and long rows of these dimers fold the inner mitochondrial membrane. You can read more on this topic in the **meandering text box**.

Meandering: Dimers of ATP Synthase Shape Mitochondrial Membranes

These days we are looking at ever larger protein complexes, due to the latest technical developments in electron microscopy. At times, these snapshots even depict proteins within their natural cellular environment. One of these protein complexes is the depicted *Fifo*-ATPase. As seen on the provided pictures, the ATPase enzyme seems not to like being alone; it prefers to appear in pairs. Several such ATPase pairs line up in a row, and together they are thought to be able to bend a membrane. ATPase dimers are therefore not a consequence of membrane curvature; they cause the membrane to fold—ATPase dimers enable the formation of stacked membranes in mitochondria or chloroplasts. Taken together, the ATPase protein not only functions as a reasonable enzyme; in addition, ATPase has a second function in shaping membranes.

And why do membrane-bending ATPases form these protein dimers in the first place? An early publication from 2008 speculated rather freely:

Probably to harvest protons as effectively as possible—the curved membrane might act like a funnel or a trap, to direct many, many protons towards the ATPase's F_O-part. The authors predicted a higher proton concentration (hence, a lowered pH value) at the curved edge of a membrane disc, than in the straight part directly next to it. In 2014 then, the local pH value in such membrane discs was measured experimentally. This was achieved by attaching pH-sensitive fluorophores to cytochrome c oxidase (complex IV) and ATPase. Surprisingly, the pH value directly at the ATPase, was about 0.3 pH units increased (!!!), not decreased, so there were fewer protons in that corner of the membrane disk, not more. One might see this still as a kind of proton trap, leading protons directly into the "F_O-mouth" of the ATPase. This trap, however, is astonishingly efficient, so that in operation the environment of the ATPase is 0.3 pH units less acidic than at the complexes of the respiratory chain—the area might be sucked empty of protons. The evolutionary truth could be completely different though. Perhaps the ATPase has simply taken on two independent functions—one is to energetically tap into a proton gradient and thus make ATP; the other is to assist in the folding of membranes in mitochondria and chloroplasts. At this moment, no one knows exactly...

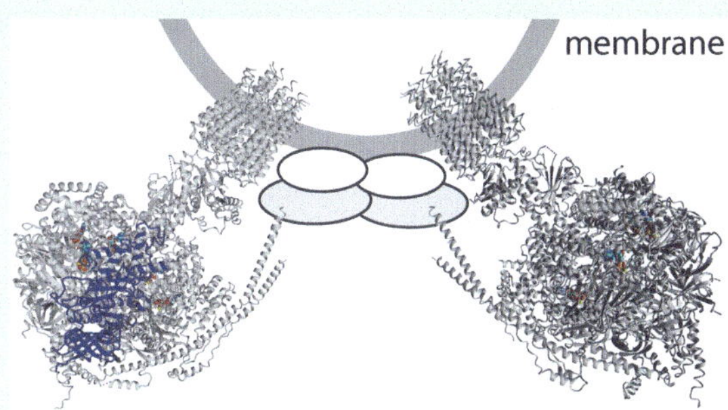

ATPase dimers bend membranes and thus might build their own funnel to capture protons better. The structure 4B2Q from the PDB protein data base by Werner Kühl brandt is visualized here. The angle of the two ATP synthase proteins to each other has been measured experimentally. The curved inner mitochondrial membrane and additional dimer-forming protein complexes are only indicated schematically.

As an analogy, pitfall traps of antlions are shown. Antlions are predatory insects, or actually the predatory caterpillars of relatively harmless insects. One rarely can see these antlions themselves. Their pitfall traps in fine sand, on the other hand, can be seen here or there. A pitfall trap can be as big as the palm of a hand; within the picture some leaves might serve as scale. The photo was taken in the South African De Hoop National Park.

The 2008 paper from Werner Kühlbrandt's group—Strauss et al. **2008**. Dimer ribbons of ATP synthase shape the inner mitochondrial membrane. *EMBO J.* 27(7): 1154ff. ► https://doi.org/10.1038/emboj.2008.35.

Exciting—using fluorescent probes, Karin Busch and colleagues are able to measure the local pH values within mitochondrial membranes—Rieger et al. **2014**. Lateral pH gradient between OXPHOS complex IV and F_OF_1-ATP-synthase in folded mitochondrial membranes. *Nat Commun.* **2014**; 5:3103. ► https://doi.org/10.1038/ncomms4103.

(Image credit: own structure visualization from ► https://doi.org/10.2210/pdb4B2Q/pdb. Antlion pit traps, South Africa, JWM, 2015)

5

Conversation Almost every biological gradient represents a source of energy; its energy content can be described by the Nernst equation. The enzyme ATP synthase accounts for the lion's share of cellular ATP synthesis, or was it the share of an antlion? The ATP synthase enzyme is driven by a proton gradient across the inner mitochondrial membrane; that proton gradient is created by the OxPhos respiratory chain.

■ Navigation

Complexes I, III and IV of the respiratory chain are electron-driven proton pumps. More on this topic in ► Sect. 6.3.

An exciting account of what Peter Mitchell had to endure for a long time. In addition to good ideas, having a pretty thick skin probably is another important ingredient for scientific success –Prebble. **2002**. Peter Mitchell and the OX PHOS WARS. *Trends Biochem. Sci.* **2002**; 27(4):209ff. ► https://doi.org/10.1016/s0968-0004(02)02059-5.

For fans: Werner Kühlbrandt writes down everything he has to say about ATP synthases. Absolutely everything—Kühlbrandt. **2019**. Structure and Mechanisms of F-Type ATP Synthases. *Annu Rev Biochem.* **2019**; 88: 515–549. ► https://doi.org/10.1146/annurev-biochem-013118-110903.

The Renaissance of Metabolism–Converting Bone-Dry Study Subjects into Modern Metabolism Research

Contents

© The Author(s), under exclusive license to Springer-Verlag GmbH, DE, part of Springer Nature 2026
J. W. Mueller, *Ultimately Understanding Biochemistry*, https://doi.org/10.1007/978-3-662-71889-6_6

In the past, we drilled into our (bio)-medical students that they must take biochemistry seriously—otherwise they would never understand the molecular basis of important diseases. At that time, probably more an attempt by the preclinical disciplines to make the slightly dusty exam material appearing somehow interesting, than anything else.

Things have changed. **Metabolic Syndrome** is a combination of high blood pressure, obesity, and type 2 diabetes. Already about a quarter of the adult world population shows several indicators of the metabolic syndrome; with a trend to increase further. Suddenly it actually is important for a doctor to understand the regulation of individual metabolic pathways. How do metabolic regulatory networks change with different diets? What is the effect of physical exertion? How do all of the above change as we age? And above all, how can we influence all these processes positively? Can we possibly even reverse aging? We have arrived at the epicenter of modern metabolism research. But first, there is some **meandering** about the greatest metabolism researcher of all time.

Meandering: The Greatest Metabolism Researcher of All Times: Otto Warburg

Otto Warburg shaped the biochemistry of the early 20th century, like none other. Some even regard him as the Greatest Biochemist of all time. It was Warburg who made biochemistry a quantitative science. Certain types of coupled enzyme reactions are called Warburg coupling, even to this very day. Warburg also made significant contributions to the biochemistry of plants. In 1931 he received a medal from Stockholm for the discovery of the enzyme cytochrome C oxidase, a component of the respiratory chain.

Today we know Warburg primarily for his discovery that cancer cells prefer to ferment sourly in the presence of oxygen. This phenomenon, known as the **Warburg effect**, persists in today's literature, even though it is discussed whether aerobic fermentation is cause or consequence of the metabolic changes in cancer cells. Nevertheless, the differences between normal cells and cancerous tumors are used for oncological imaging these days. In addition, anti-metabolites or inhibitors of basal metabolic reactions are successfully used in chemotherapy.

In times of very tight budgets, Warburg's application was funded in full. Irrespective of this, he probably wasn't particularly pleasant in daily life. He pursued his research with an intensity and unconditionality that gave many employees a hard time. Warburg never married. He lived together with his chauffeur, more or less openly gay. The photo shows Warburg together with his poodle *Bärchen*, which probably best translates to "Teddy Bear". Warburg was of Jewish decent. Despite all this, he continued his research in Berlin during the Second World War almost without interruptions. What a biochemist.

Otto Warburg was a very successful and highly respected scientist. His "somewhat succinct" 1921 research application to what later became the

German Research Council DFG is legendary. A reconstruction based on a detailed description from H. Krebs, is shown alongside Warburg's portrait, saying "Application. I need 10,000 (ten thousand) marks. Otto Warburg."

Dr. Otto Warburg

Antrag

Ich benötige 10 000 (zehntausend) Mark

Otto Warburg

Without a doubt, the life and work of Otto Warburg have long been first-class material for a novel. Finally, it has happened. Sam Apple published his book on Warburg in 2021—Apple S. **2021**. Ravenous—Otto Warburg, the Nazis, and the Search for the Cancer-Diet Connection. Published by Liveright Publishing.

For those who prefer something more scientific-historical, here is a decent account of Otto Warburg's life and his scientific achievements—Koppenol et al. **2011**. Otto Warburg's contributions to current concepts of cancer metabolism. *Nat Rev Cancer*. 11(5): 325ff. ► https://doi.org/10.1038/nrc3038.

There are several recent review articles that transport the Warburg effect into the modern age, for example this one—Potter et al. **2016**. The Warburg effect: 80 years on. *Biochem Soc Trans*. 44(5): 1499ff. Review. ► https://doi.org/10.1042/BST20160094.

(Photo: Suse Byk, around 1925, Archive of the Max Planck Society, Berlin. Reprint with permission; Warburg's grant application, adapted from Koppenol et al. Re-use with permission from Springer Nature)

6.1 The London Underground Network and Emergence at Various Layers

To be a good biochemist or metabolism researcher, you don't need to know that TIM, our beloved triose-phosphate isomerase enzyme, catalyzes the isomerization of GAP and DHAP via an intermediate diol step. It is probably enough to know that, in the cell, GAP and DHAP are almost equivalent as long as TIM is around. Similar thinking applies to the London Underground network (■ Fig. 6.1): For some, it is enough to know that you can change between different underground lines at Euston Station—it is a bit complicated and involves a lot of walking. For others, it is important that you can board the high-speed train to Birmingham at this station. Probably superfluous knowledge of unnecessary details might be the fact that just next to the station, there is a traditional pub, offering a large selection of freshly draught beers and also quite an acceptable menu of affordable food. One can live without knowing this.

For this chapter, we want to stick to the main lines of the network, the major transfer points or nodes in the network, as well as a few important remarks on operational oddities. Such announcements may include the label "one-way road" or "Station closed on Saturday, due to farmers market". To explore some side lines of the network, I recommend contacting other biochemical literature, thinking of the biosynthetic pathways of different amino acids or nucleobases, of the making of cofactors and other specialists, and a lot of degradation pathways. Unfortunately, we will not mention this small "café", that incredibly interesting "boutique", the newly opened "gallery" or the traditional "corner pub" in this book. So much to tell, but after all, it is an overview, I'm presenting to you here.

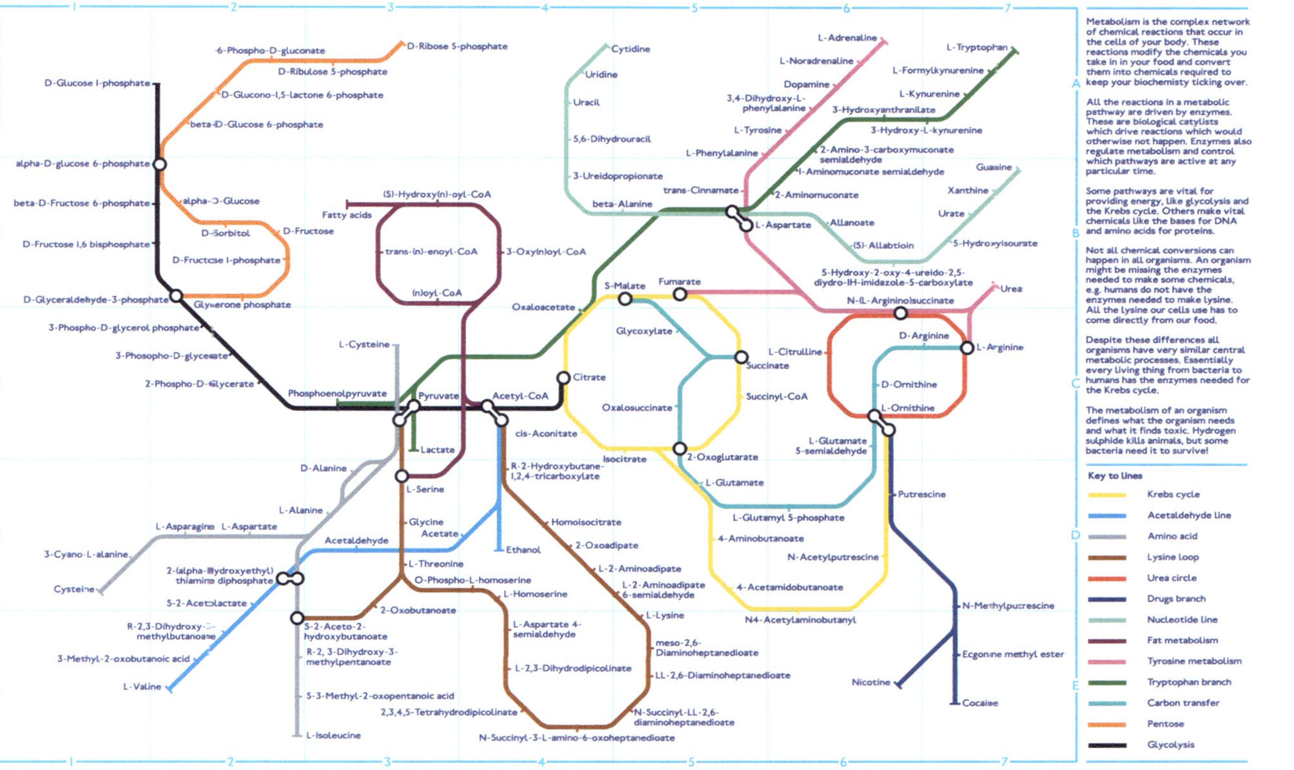

Fig. 6.1 **The central metabolism of the cell in the style of the London Underground map**. This map invites you to browse through cellular metabolism in a unique way. Perhaps it also provides some food for thought. If you like, you can match colors of pathways with the London Underground map. Maybe not coincidentally, the color black is shared between glycolysis and the Northern Line—both ancient and venerable pathways. (Image credit: Richard Wheeler, 2019, Reprint with permission)

6.2 Glycolysis, Fermentation, Glycogen and Gluconeogenesis—A Quick Run-Through

We enter the glycolysis line with grape sugar, also called glucose. Glucose is absorbed by the small intestine from our food and phosphorylated instantly by hexokinase in almost all cells. BANG, glucose thus has acquired a negative charge and can no longer easily leave the cell. As almost always during metabolism, something special happens in the liver. Liver has a different enzyme at hand to phosphorylate glucose—liver's very own glucokinase enzyme. **Glucokinase and hexokinase** are a prime example of an unequal pair of related enzymes. Both enzymes use ATP to produce glucose-6-phosphate; and this phosphorylated sugar Glc-6-P represents the first central transfer point in our metabolism subway network. More about this rivalry between cousins in the following **meandering text box**.

Meandering: Hexokinase and Glucokinase, Two Unequal Cousins

You can refer to two enzymes that catalyze the same reaction, as isoenzyme. Here, "iso" is the Greek word for equal. Human glucokinase and the catalytic part of hexokinase are pretty similar, 54% of their amino acids are identical.

Hexokinase is found in high concentrations in almost every tissue. This enzyme works decently at quite low amounts of glucose; hexokinase becomes quickly overwhelmed however, when glucose is available in abundance. Biochemists speak of a low K_M constant and a low v_{max} value (we have already discussed these parameters in ► Sect. 5.1). The liver offers a special treatment for glucose. There, the enzyme glucokinase is found, instead of hexokinase. Although somewhat sluggish at low glucose concentrations, glucokinase really gets going, when more sugar becomes available, and then hardly ever stops. The biochemist would attest glucokinase to have a higher K_M constant and a higher v_{max} value.

Every tissue that is not liver is grateful for every tiny bit of glucose absorbed. You never know, but blood glucose levels could drop unexpectedly. So, probably better to absorb and immediately phosphorylate whatever glucose you can get—biochemical "tagging" or "locking" in a way. At some point however, the need is met—peripheral tissues are saturated—one should not be too greedy. Liver is quite different—it is the major sugar producer (keyword **gluconeogenesis**) and also one of the major sugar storages (keyword **glycogen**). After a carbohydrate-rich meal (for example, spaghetti with a yummy tomato sauce), liver literally "soaks" up glucose—just like a sponge. Because it is good for us.

Physiologically, the division of labor between the two enzymes glucokinase and hexokinase makes perfect sense. In addition, these enzymes are a perfect example of metabolic regulation that is firmly engrained into the structure and functioning of the proteins involved.

Small Detour—Glycogen

At times, there is so much glucose around on the biochemical tracks that parts of it is sent on a small detour. Liver and muscle are our major glucose storages, they can either store or release glucose, due to the body's needs. Liver can do this by converting glucose into **glycogen**—muscle tissue and kidney, to a certain extent, can do the same. Glycogen is the animal equivalent of plant starch—referred to as liver starch, in the past. Starch and glycogen are long chains of glucose, that are more or less branched like trees. While the unbranched starting point is barely accessible—it holds the structure together, glucose can be quickly broken down from the branched end. Well, in reality, that branched end is very, very many ends. The trick is: a glycogen particle with up to 50,000 glucose units osmotically counts like a single molecule—attracting only tiny amounts of water. A liver cell could not store the same amount of free glucose—never. The cell would burst osmotically, due to all the water that the glucose brings with it. Stuffing itself with a few hundred glycogen particles instead, is perfectly possible however.

Making and breaking—the synthesis and breakdown of glycogen are strictly regulated hormonally. The storage of chemical energy in the form of a glycogen granule is incredibly energy efficient—**napkin XIII** approximates how efficient this process actually can be. The most remarkable step in glycogen is the actual cleavage of a glucose unit, which is accomplished by a very unusual enzyme, using unusual biochemistry. Glycogen phosphorylase binds the cofactor pyridoxal phosphate "upside down". The cofactor no longer presents an aldehyde group, but the phosphate group instead. It is this phosphate that is pressed into the bond between the terminal and second-last glucose unit. Ta-dah, glucose-1-phosphate is created immediately—without any ATP consumption. This phosphorylated sugar Glc-1-P then only has to be converted into glucose-6-phosphate, to (re-) enter cellular metabolism.

Napkin XIII: Glycogen as Highly Efficient Energy Storage

A glycogen storage granule is a delicate thing. Similar to a huge tree, it only has one trunk, but almost infinite numbers of ends on the side of the branches. How such a particle exactly forms is explained in other textbooks. The following calculation applies not for building the whole granule; it is specifically for a molecule of glucose-6-phosphate, which is attached to one of these many ends and detached again later:

Attaching a glucose unit

(1) Glc-6-P → Glc-1-P – enzyme: phosphoglucomutase; energetic cost: almost none

(2) Glc-1-P + UTP → Glc-1-UDP + PP_i – enzyme: UDP-glucose-pyrophosphorylase; cost: cleavage of a single phospho-phospho bond

(2+) PP_i → 2 P_i – enzyme: pyrophosphatase; cost: cleavage of another phospho-phospho bond

(3) Glc-1-UDP + $(\text{glycogen})_n$ → $(\text{glycogen})_{n+1}$ + UDP – enzyme: glycogen-synthase; cost: cleavage of a phospho-carbohydrate bond

Detaching a glucose unit

(4) $(\text{glycogen})_{n+1}$ + an ordinary phosphate ion → $(\text{glycogen})_n$ + Glc-1-P – enzyme: glycogen-phosphorylase; cost: none

(5) Glc-1-P → Glc-6-P – enzyme: phosphoglucomutase; energetic cost: none

All in all, the cell has to invest the energetic equivalent of cleaving two ATP to ADP to temporarily store a molecule of glucose-6-phosphate, including all fixating and freeing steps. If this very Glc-6-P molecule later is fully oxidized, generating about 34 ATP molecules from ADP and inorganic phosphate, then we would have a storage efficiency of 94% for this biochemical battery; pretty good, isn't it… In this example, we only considered glucose that was attached to the very non-reducing end of the glycogen particle. We ignored branching and de-branching enzymes. Of
6 course, these are needed to build the glycogen granule in the first place. Because of the tree-like or dendritic structure of the glycogen granule, however, this napkin's approximation applies to almost half **of all** glucose molecules within the glycogen particle.

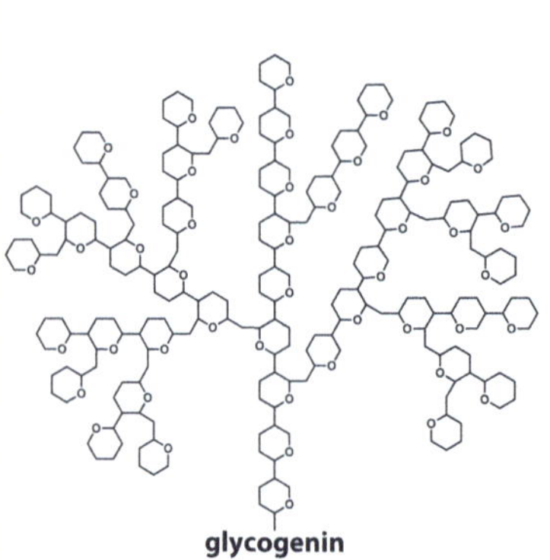

(Image credit: Diagram, JWM 2022; Tree, © Eduard Vladimirovich stock.adobe.com)

Glycogen is just great. You can pack up to 50,000 glucose units into a glycogen particle, to be stored almost osmotically neutral. The dendritic structure of glycogen resembles a tree. Almost half of all glucose residues are located in the outermost shell; hence they are accessible with ease. So, you would get VERY many glucose units out of glucose, following the approximation above. What a great storage form for glucose! The body can draw energy from glycogen extremely quickly, and—in contrast to the combustion of free fatty acids—energy from glycogen could even be made without oxygen.

The Glycolysis Line Runs Every Three Minutes

In ◘ Fig. 6.1, the glycolysis line is shown in black in the metabolism diagram—black also is the color of the Northern Line on the London Underground map. The Northern Line is the oldest subway line of the network. Similarly, almost all cells contain much of all the glycolytic enzymes in abundance. Being around for long, and in big amounts, are perfect prerequisites for evolution to put things to additional uses—for example, the enzyme enolase serves an additional purpose in the lens of our eye as a structural protein. Biochemically speaking, the high concentration of glycolytic enzymes means that they are almost permanently underemployed. An apt comparison probably could be an almost empty highway. In other words, at any time, the cell can send a larger amount of glucose-6-phosphate down the glycolysis highway, all that glucose would be metabolized almost immediately. This is very important when energy is needed quickly.

All glycolysis intermediates are phosphorylated so that the glycolytic enzymes involved can bind them tightly and specifically. Energetically, glycolysis is a roller coaster ride, initially energy has to be put in, before energy comes out again. The local strategy of glycolytic energy generation is always the same: to make high-energy molecules from lame molecules and then to generate ATP from those high-energy molecules. Affectionately, this strategy is called substrate chain phosphorylation.

The ATP-consuming steps within glycolysis are the production of glucose-6-phosphate and the phosphorylation of fructose-6-phosphate to fructose-1,6-diphosphate (also called fructose-1,6-bisphosphate). The energy-yielding steps are associated with 1,3-bis-phosphoglycerate (1,3-BPG) and with peppy phospho-enol-pyruvate (PEP). In ▶ Sect. 5.3 we already have tried to understand why molecules like PEP and 1,3-BPG house more chemical power than ATP.

By the terminal stop of the glycolysis line, which is pyruvate, we have invested two ATP per glucose unit, but got four ATPs out of it. That glucose has given rise to two molecules of pyruvate—unlike all the intermediates of glycolysis, pyruvate is no longer phosphorylated. Finally, we have "harvested"some electrons and stored them in NADH. The whole process went pretty quickly and without delays. With enough sugar, life had never been sweeter. If, only, we could get rid of all these electrons.

The Short-Haul Ticket: Lactic acid Fermentation Provides Oxygen on Credit

Proper oxidizing stuff takes time, as we will see in the next chapter. If we were in a hurry, however, our metabolic subway ride is about to end, not too far away from pyruvate. The enzyme lactate dehydrogenase (LDH) takes the practically worthless

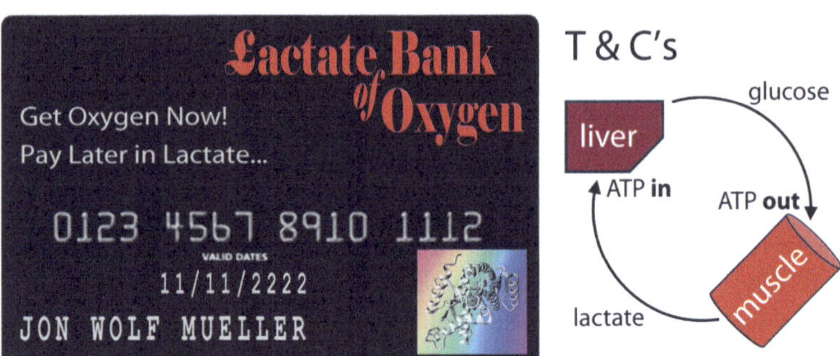

Fig. 6.2 The lactate credit card buys ATP at any time, no oxygen required. Glycolysis and lactate fermentation can generate ATP in the muscle, even in the absence of oxygen! Handy, when things need to go quickly—imagine you run to catch your bus, your train, or your plane, okay, you can add "running away from a lion". If you want, you accumulate an oxygen dept in your muscles. In the form of lactate, this "oxygen credit" is then transported to other organs—mainly to the liver. When enough oxygen is available again, lactate laboriously is assembled back into glucose. This shuttling of lactate back and forth is also called the Cori cycle. (Image info: JWM, 2021; own structure visualization of human LDH, ► https://doi.org/10.2210/pdb1I0Z/pdb; schematic, JWM, 2025)

6

pyruvate, and those annoying electrons stored in NADH. LDH recycles all that into lactate and NAD. The NAD cofactor thus is free for another round of glycolysis. Leaving lactate to accumulate for a while (Fig. 6.2). What happens then, depends on where these processes take place.

We humans rely on fast energy from glycolysis and waste recycling via lactic acid fermentation, whenever our muscles enter the "anaerobic" zone. It is then when the muscle needs to perform more than it can power through aerobic respiration. As a coping mechanism, muscles generate ATP through glycolysis and fermentation, only to accumulate large amounts of lactic acid, in the form of lactate. Lactate is then released into the blood, and taken up by the liver, to be processed further, once there is opportunity for it. If you allow me to put it like this, muscle just has taken an oxygen credit from the liver, and lactate is accepted as a means of payment. Like in real life, the lactate line of credit will come to an end at some point. Lactate accumulating in the muscle will ultimately acidify the tissue around it, while considerable muscle activity endures. Be it caused by acidification or microscopically small fiber ruptures, we most likely will have thoroughly aching muscles about 24–72 h after that intense "exercise".

Cheesy Rapid lactate fermentation keeps the muscle running even in the absence of oxygen. Some microorganisms are truly silent masters of various types of fermentation. They don't aspire towards rapid ATP production. Slowly, but steadily, they work towards transforming food instead. This topic is addressed in a **meandering text box** on connected learning, even if some of the connections were somewhat indirectly.

Meandering: Connected Learning of Biochemical Fermentation Types: A Personal Account

Back in the days of my study, I was studying for a major biochemistry exam. I did so together with a fellow student of mine. We crammed glycolysis and fermentation for that test. A good while into the session, I opened a bottle of beer and said: "actually quite handy, the metabolic pathway we are just studying". Truly confused, partially surprised, my friend asked back to me… "Why?"… The possibility that that substance "ethanol" from the textbook and the "alcohol" contained in our beverage could share any kind of relationship, maybe even the chemical fact that the two were identical down to the atom, had not crossed his mind until then…

Many years later, this story still astounds me—how can people absorb facts and figures, seemingly WITHOUT cross-referencing them with their own lived experience. Back to fermentation, you already could guess that discussing glycolysis and fermentation might be easier for some of us with a beer in hand. Discussing lactic acid fermentation, might be paired with a mild cheese and a good wine. For fans: Any elaboration of propionate fermentation that follows lactate fermentation, could go well with a strong Swiss Emmental cheese and perhaps an elegant fruity sherry.

We probably have eaten enough cheese by now. Even if this strategy cannot be applied to every course of study of life sciences, at any given time, I personally would try to establish some sort of a personal connection to the learning objectives, always. This approach to learning hopefully makes it easier to learn and understand whatever subject, when you are involved in it also subconsciously—whole-heartedly, or whole-guttedly, as seen here.

Further Reading

David Goodsell provides a great overview of glycolysis in the great PDB—101 education section of the Protein Data Bank, the PDB—Goodsell. **2004**. Glycolytic Enzymes. Molecule of the Month. RCSB PDB-101, Educational portal of the PDB. ▶ https://doi.org/10.2210/rcsb_pdb/mom_2004_2 [Accessed 6.6.2022]

6.3 Attention: The Road from Pyruvate to Acetyl-CoA is ONE-WAY Only; and two more pathways to discuss…

Should we not engage in any kind of fermentation, the next stop after pyruvate is an acetic acid residue, which is bound to coenzyme A. A bit of electron shuffling and harvesting, kicking out of CO_2, and bringing in coenzyme A. Doesn't sound particularly difficult, does it? But it is. This reaction is catalyzed in five partial steps by one of the largest enzyme complexes roaming around in the cell—the pyruvate dehydrogenase complex, a gigantic machinery of about 9 mega-Daltons. That's almost the weight of THREE ribosomes at once! This complex contains a handful of enzymatic cofactors: TPP, lipoic acid, coenzyme A, FAD and NAD. Reminder: Since these five cofactors, and many others alike, are derived from various vitamins, please be reminded to eat vitamin-rich food plenty so that your metabolism always works well.

Yeasts don't particularly like that large PDH enzyme complex, the pyruvate dehydrogenase. Yeasts prefer to employ the somewhat more compact pyruvate decarboxylase, PDC, instead. PDC is a fabulous enzyme. With PDC, yeasts produce ethanol and carbon dioxide from pyruvate—an extremely useful reaction for humans, both for making pizza or baking bread, as well as for the production of beer and wine. By the way, all foods where yeasts are involved, such as bread and beer, contain abundant amounts of most components of the vitamin B complex—good for your metabolism.

Caution Our human metabolism knows the connection between pyruvate and acetyl-CoA only as a ONE-WAY road. Meaning that our liver (well, who else) can build glucose again from almost all carbon-C3 bodies, including pyruvate. Once converted to the C2 body acetyl-CoA however, those carbons will most likely end up in fat; at least in human metabolism, the way back to glucose and other sugars is closed.

Circle Line: The Citric Acid Cycle

We have arrived at the center of human metabolism, the citric acid cycle—a mega-hub. You are completely free to choose the most suitable traffic routes analogy, here. We can remain travelling along the Circle Line—maybe a good choice since traffic load has been shared with the new Elizabeth Line in 2022. Instead, we also could mentally switch to the car and enter a gigantic roundabout—why not think of the huge roundabout around the Arc de Triomphe in Paris, France.

Back to the citric acid cycle: It takes eight clever biochemical transformations to completely break down the small acetate appendix in acetyl-CoA into carbon dioxide, most literally ending up in smoke. In a full round of this metabolic carousel, a lot of electrons are harvested and even a GTP molecule is produced, thanks to cunning substrate-level phosphorylation chemistry, again. Some say somewhat sarcastically that the actual Krebs cycle consists of learning it and forgetting it soon after. Maybe there is some truth in it.

Could one learn the Krebs cycle in a more sustainable fashion. On top of really interesting biochemistry, well, and maybe geo-biochemistry, to be explored in ▶ Sect. 6.7, there is exquisite regulation around the Krebs cycle. Then there are versions of the cycle, mainly in bacteria, that go the other way around. And finally, there are replenishing reactions; there are shortcuts; and lots of branching-offs into other biosynthesis pathways. Your learning strategy might best be informed on what you want to learn for; be it just mastering a classroom test or understanding the effect of some dietary intervention.

The core reactions of the Krebs cycle are highly conserved in basically all forms of life. Contrary to all the phosphorylated metabolites of glycolysis, not a single one in the Krebs cycle carries a phosphate group. You may wonder how all these reactions could have arisen randomly and simultaneously? You are not alone, as people in the past have postulated an appeal to magic, when it comes to the Krebs cycle. We will compare the Krebs cycle with a magical roundabout, and will give possible answers about the origin of the Krebs cycle in ▸ Sect. 6.7.

The Oxidative Phosphorylation—OxPhos Hovers Above It All

Next stop is the Oxidative Phosphorylation, or OxPhos for short. Should you still want to stay in the realm of traffic analogies, OxPhos would be a very, very strange subway line. OxPhos no longer runs on a carbon basis, instead OxPhos mainly pumps electrons, and this indirectly also leads to the pumping of protons. If you still are up for an analogy from the London transportation network, OxPhos probably can be compared to the scheduled river ferries between Putney Pier in the West and Barking Riverside Pier in the East.

The OxPhos machinery consists of five super complexes, creatively named CI, CII, CIII, CIV and CV, where the "C" stands for "complex", so their actual names are I to V. Complexes I, III and IV are all electron-driven proton pumps. Complex II is a kind of adapter complex that can squeeze medium-energy electrons into the respiratory chain, which Complex I would not accept. Complex IV finally uses oxygen to get rid of the electrons, it produces water in this way. All the complexes I–IV work hard to establish a proton gradient amongst the inner proton-tight mitochondrial membrane. Complex V is the ATPase feeding on this proton gradient across the inner mitochondrial membrane to generate ATP. We introduced it in ▸ Sect. 5.7. The electrons are gone by now. A good part of the energy is stored in the form of ATP. Some warmth was released. Everyone is happy.

Cascade Not far away from the Krebs cycle, a C3 body is converted into acetyl-CoA quite tediously. This conversion is a one-way street in humans—we cannot make glucose out of acetyl-CoA anymore. This very acetyl-CoA is completely converted into CO_2 fumes, some high-energy molecules, and a lot of electron-charged electric storage units. Very complex mitochondrial respiratory complexes are electron-driven proton pumps. They create a proton gradient over the inner mitochondrial membrane, by ultimately bumping the electrons on oxygen, producing water. … and then there is lollipop-ATPase—powered by the proton gradient, it finally generates ATP, lots of it. Done. And by the way, if you feel your summary of OxPhos is smarter than mine, please get in touch.

■ **For Further Amazement**

Just a simple enzyme, but a gigantic one, well, actually it's not so simple: Focus on the huge pyruvate dehydrogenase complex—Patel et al. **2014**. The pyruvate dehydrogenase complexes: structure-based function and regulation. *J Biol Chem*; 289(24):16615–23. Review. ▶ https://doi.org/10.1074/jbc.R114.563148.

6.4 The Pentose Phosphate Pathway Protects Us from Harmful Reactive Oxygen Species

Since we crammed all the other metabolic pathways into one or two compact chapters, one might ask: Why does the pentose phosphate pathway get a chapter all to itself? The simple answer is: Because it deserves it. While we compared glycolysis to a fully developed motorway or the eminent Northern Line of the London subway network, the pentose phosphate pathway more resembles a number of winding paths, perhaps a tram network in a residential and business district, where you can get many things done, or none.

The pentose phosphate pathway, affectionately also called PPP, is an alternative way for the body to burn or to oxidize glucose. The first oxidation step occurs at C-atom No. 1. During this reaction, the hexose ring remains closed—the glucose molecule remains in ring form for now. The aldehyde group, which was disguised in a half-acetal bond, transitions to a carboxylic acid. At the same time, the type of bond changes—an internal ester is formed, a lactone. That lactone then is cracked by a specialized lactonase enzyme, in this second reaction step the hexose ring is opened. The product 6-phospho-gluconate is stretched out now, ready for the third reaction. The oxidation takes place at C-atom No. 3, allowing to kick out the C1 unit as carbon dioxide fumes.

This pattern of two oxidation reactions is—more or less—already known from other metabolic pathways. The big difference is that the two oxidizing enzymes at work here—glucose-6-phosphate dehydrogenase and 6-phosphogluconate dehydrogenase—use NAD**P** as a cofactor and not NAD; in this way NADPH is produced, not NADH. Therefore, the glucose is burned here for a good cause, not mere energy production, but to generate reduction equivalents for biosynthesis pathways. NADPH finds alternative use for the defense against harmful reactive oxygen species.

The leftover is ribulose-5-phosphate. In dizzying precision, this C5 sugar is amalgamated with some other sugar, then severed again, back and forth, and so on and so forth. All these "knitting-like" reactions are realized by just two enzymes, by the TPP-dependent transketolase and the "cofactor-free" transaldolase—C2- and C3-bodies are transferred between different phosphorylated sugars; and all these reactions occur almost at equilibrium conditions. Yes, a bit like in a knitting pattern—two to the left, three to the right…

It is not that easy to understand why transketolase and transaldolase work so differently. Both enzymes focus on the same keto group, be it for a C2- as well a C3-transfer. To get either reaction done, an emerging negative charge at the carbon atom of the keto-group must be stabilized. Transketolase uses thiamine pyrophosphate (TPP), a cofactor that can never have enough electrons and therefore easily forms electron-pulling carbon-carbon bonds.

Transaldolase, on the other hand, puts forth a lysine side chain with a free amine group. That lysine engages the sugar substrate in a Schiff base, which is a C=N double bond formed between the lysine amine group and the sugar carbonyl group, formed while a molecule of water is kicked out. TPP or Schiff base, subtle electronic differences between these two approaches and a number of elaborate protonation and deprotonation events make the one reaction to be a C2-transfer, while the other one transfers a C3-body. How this all will end, depends on the body's needs.

One outcome of all this is that ribose-5-phosphate might be produced—very handy if you want to synthesize **nucleotides** for RNA, for DNA, or for whatever. Erythrose-4-phosphate is another end product needed for the biosynthesis of **aromatic amino acids**. Another outcome could be the congruent production of glyceraldehyde-3-phosphate and fructose-6-phosphate, both of which could seamlessly re-enter glycolysis and gluconeogenesis; only to run through the pentose phosphate pathway again—this way, an entire glucose molecule can be metabolized to **NADPH and CO_2**. This last version of the PPP is particularly common in tissues that have to deal with increased amounts of reactive oxygen species—such as red blood cells, which transport the "hazardous material" oxygen. The "dirty" steroid factories of the adrenal gland are another tissue with a lot of PPP activity.

Community The pentose phosphate pathway (PPP) allows the body to make use of glucose in unique ways. The PPP can run very flexibly. Products can be various C4- or C5-sugar units. Reduction equivalents in the form of NADPH are also produced. NADPH then can be used for biosyntheses or for the defense against reactive oxygen species. Transportation-wise, the PPP is somewhat more leisurely than glycolysis.

Further Reading

Kai Tittmann has always liked thiamine pyrophosphate-containing enzymes. Here is his portrait of transaldolase and transketolase—Tittmann. 2014. Sweet siblings with different faces: the mechanisms of FBP and F6P aldolase, transaldolase, transketolase and phosphoketolase revisited in light of recent structural data. *Bioorg Chem*. **2014**; 57: 263–280. Review. ▶ https://doi.org/10.1016/j.bioorg.2014.09.001.

When you think you know it all, this happens. Cells can put accumulating redox equivalents on unsaturated fatty acids, not just by making lactate—Kim et al. **2019**. Polyunsaturated Fatty Acid Desaturation Is a Mechanism for Glycolytic NAD+ Recycling. *Cell Metab*. 29(4):856–870.e7. ▶ https://doi.org/10.1016/j.cmet.2018.12.023.

6.5 Burning Fatty Acids, and Re-making Them

Fats and oils themselves can be tasty, yummy, or savory. In addition, they serve as carriers or solvents of a lot of fat-loving, lipophilic, flavory metabolites. Please think of peppermint oil now, or chili oil, or perhaps just the oil from roasted sesame seeds.

Above all, fats are stores of energy. Thinking of the noticeable fat accumulation in my abdomen, my big belly, it is quite impressive what a substantial energy reservoir I have there. We don't need to do any calculations on a napkin here—most realistically, I would die of thirst before I would starve to death. So, this chapter is about how fatty acids are burned in the cell, for energy generation; and also, how they are made, for energy storage.

Our mitochondria are where fatty acids are burned, in a process affectionately called **β-oxidation**. A strange term, which has its origin in the traditional α-β-naming of carboxylic acids. For merely historical reasons, biochemists do not count the carbon atom of the carboxyl group itself—it would have been carbon number one, otherwise. The next carbon atom then is called α and the following one is referred to as β. That's just how it is. For β-oxidation to occur, it is important what happens at this very β-carbon atom, which in fact is the third C-atom of a fatty acid. This β-carbon is oxidized step-by-step.

Chemically, the β constellation is something special: The Cα-atom is usually quite acidic due to its proximity to the carboxyl group and theoretically more inclined to give away a proton, than other hydrogen-carrying carbons. As there will be a double-bond between Cα and Cβ at some point, affectionally called an "allyl", the β-carbon is getting an increasingly positive twist. One round of β-oxidation—without the cleavage step, yet—creates an activated fatty acid with the β-carbon now oxidized to a keto group. This β-keto-carbon is then attacked by the thiol group of a coenzyme A molecule, splitting it at the very Cβ atom. The result is a fatty acid shortened by two carbon atoms, some electrons harvested, and a molecule of acetyl-CoA.

For a long time, metabolism researchers thought that the **synthesis of fatty acids** was the exact reversal of β-oxidation. That's true, more or less. It just is that almost EVERY step is different... And then there is this little issue with fatty acid synthase, the enzyme-in-charge. Fatty acid synthases, who could have guessed, are actually a little bit complicated. Bacteria employ a group of distinct proteins to make up their fatty acid synthase complex. The individual enzymatic activities are neatly parceled out into several polypeptide chains. All types of fatty acid synthase machines presented here, share one thing—all of them employ a flexible arm where carbon unites are moved back and forth; none of these machines would release intermediates of this pathway into the cell.

In contrast to bacteria, we, let's say "all mammals" this time, have all catalytic activities needed for making saturated fatty acids, nicely lined up on a single huge polypeptide. In classical textbooks you may find this enzyme mega-complex sche-

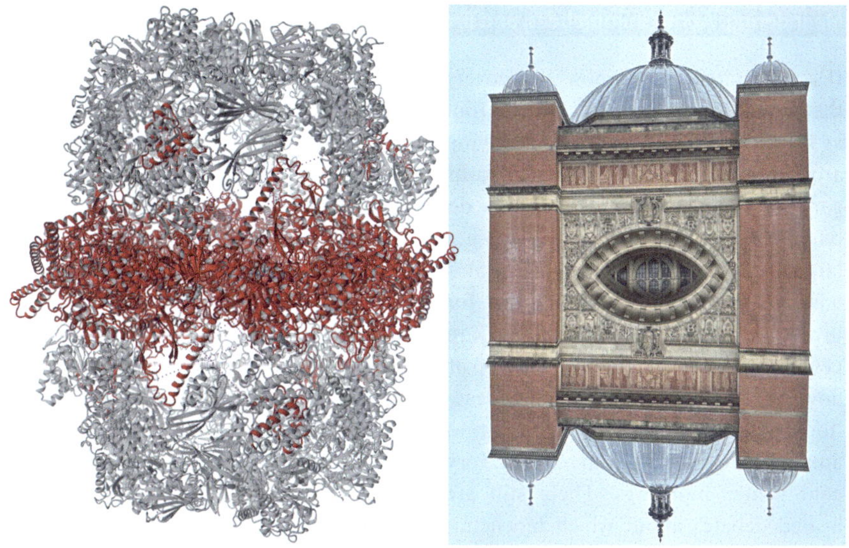

■ **Fig. 6.3** **Fatty acid synthase always was something special**. A schematic representation of a cryo-electron-microscopic structure of yeast fatty acid synthase, shown at the left. What a beauty—a heteromeric 12-mer, consisting of 6 α- and 6 β-chains. The whole complex comes with a molecular weight of about 2.65 Mega-Daltons. It reminds me of the domes that we have at the center of our Birmingham campus in Edgbaston. One might think of a double dome as a gigantic model of fatty acid synthase, shown at the **right**. (Picture credit: structural representation of ▶ https://doi.org/10.2210/pdb6U5U/pdb, JWM, 2025; domes at Chancellor's Court, JWM, 2023)

matically drawn as a yin-yang symbol (not shown here, though), suggesting that the complex works as a dimer. In reality, this big thing forms something of an upper and a lower part of a hamburger bun, with catalysis happening somewhere in between—somewhere where you would expect the burger patty.

Finally, there is another form of fatty acid synthases, found in fungi, in mushrooms, in molds. Their fatty acid synthase often consists of several copies of two different polypeptides—they form a humongously large, but elegant mega-complex. I can only describe these multifaceted factories of fatty acids as fascinating. ■ Figure 6.3 shows the dome-like fatty acid synthase from yeast, alongside with some domes from our campus in Birmingham.

And then there's the problem with storing fatty acids, vaguely similar to storing sugars. Free fatty acids are only marginally soluble in water; they also tend to stick to all possible surfaces. For sending them through our body's circulation, fatty acids are neatly bound to albumin as a transport device. For medium- and long-term storage, fatty acids are converted to neutral fats. These then are stored in different types of adipose tissue in the body, which are briefly introduced in the following **meandering** section.

6

Meandering: About Bumblebees, and Brown, and White, and Beige Adipose Tissue

It's cold outside and snow still covers the landscape. But at temperatures from as little as 2 °C, **bumblebees** are flying around, exploring the lavender in your garden, among other plants. How do bumblebees do that? Of course, they are larger and heavier than bees. They also have a much denser coat of hair for insulation. In addition, bumblebees can generate heat—seemingly out of nowhere—by running biochemical reactions in a cycle, somewhat similar to auxiliary heating in your car, at very cold winter mornings. There still are heated debates about which biochemical reactions exactly contribute to non-shivering thermogenesis. The elevated simultaneous activity of the enzymes phosphofructokinase and fructose-1,6-bisphosphatase used to be a hot candidate.

Humans can generate heat without a shiver, mostly in brown adipose tissue. We all love **brown adipose tissue**. About 5% of the total body weight of a newborn consists of it. Brown adipose tissue helps babies to survive. It protects the offspring from cooling down and every dad or proud aunt can report about the heating, a baby can provide. In brown adipose tissue, a lot of the protein UCP1 can be found—UCP1's other name is thermogenin. This protein decouples the respiratory chain and ATP synthesis. Mitochondrial proton gradient protons just can flow back into the mitochondrial matrix; ATPase runs empty here… and the OxPhos fun can start again—heat is generated in the process.

Unfortunately, adult humans no longer have a lot of brown adipose tissue. What many of us have more than enough of is **white adipose tissue**, instead. Encouraging are findings that white fat can be "browned"—resulting in **beige fat**, which to a certain degree can also generate heat. Beige fat also contains some amounts of the UCP1/thermogenin protein. But generating heat there is not that simple. It seems that energy is a VERY precious commodity that cannot be eliminated by a single short circuit. Much more, in beige fat, several switches in metabolism are turned so that we feel comfortably warm.

▪ Further Reading

A provocative article on why it doesn't always have to be UCP1/Thermogenin—Brownstein et al. **2022**. ATP-consuming futile cycles as energy dissipating mechanisms to counteract obesity. *Rev Endocr Metab Disord.* 23(1): 121–131. Review. ▶ https://doi.org/10.1007/s11154-021-09690-w.

For the humble bumblebee at least, this study from 2017 see glycerol-3-phosphate dehydrogenase at the forefront of biochemical auxiliary heating—Masson et al. **2017**. Mitochondrial glycerol 3-phosphate facilitates bumblebee preflight thermogenesis. *Scientific Reports.* 7(1): 13107. ▶ https://doi.org/10.1038/s41598-017-13454-5.

Compact The breakdown of a fatty acid into C2 chunks—acetyl-CoA—works flawlessly. Every now and then, however, our metabolism has to exert itself a bit with branched fatty acids and also with those that contain an odd number of carbon atoms. The synthesis of fatty acids runs in principle exactly in the opposite direction, only each individual step is different. The C2 units are activated by carboxylation; NADPH is always used for reductions. Overall, fatty acids are made in the cytosol by a monster protein complex that is almost as large as the ribosome is. In contrast, β-oxidation takes place in the mitochondrion.

■ **For Fact Checking**

We have discussed fats already in ▶ Sect. 2.3.

Probably a very noteworthy source for the structural elucidation of the fatty acid synthase by the working group of Nenad Ban, a former coworker of the crystallographer Tom Steitz. Some people have commented in jest that Ban only worked on this protein because he erroneously beheld fatty acid synthase as the ribosomal fraction in some purification, simply because of their similar size—Maier et al, **2010**. Structure and function of eukaryotic fatty acid synthases. *Q Rev Biophys*; 43(3): 373–422. Review. ▶ https://doi.org/10.1017/S0033583510000156.

And **two** new whopper papers, published back-to-back, describing how fatty acid synthase works in real time. The first **ONE**—Choi et al. **2025**. Structural dynamics of human fatty acid synthase in the condensing cycle. *Nature*. ▶ https://doi.org/10.1038/s41586-025-08782-w. The **SECOND** one—Schultz et al. **2025**. Snapshots of acyl carrier protein shuttling in human fatty acid synthase. *Nature*. ▶ https://doi.org/10.1038/s41586-025-08587-x.

6.6 Regulation of Metabolic Pathways and Metabolite Networks

Autonomous and local regulation is like a thermostat to ensure that a metabolic pathway does not have too much, nor too little throughput. A good example is the regulation of the pentose phosphate pathway: the rate-determining step is the oxidation of glucose-6-phosphate. The responsible enzyme Glc-6-P-dehydrogenase is conveniently activated by NADP; even better is that this enzyme is strongly inhibited by NADPH. As soon as NADPH is consumed, it is reproduced; NADPH *on-demand* so to speak. Should NADPH accumulate however, the enzyme Glc-6-P-DH switches off.

RNA thermometers are a fantastic way of sensing temperature fluctuations in the microenvironment. In bacteria and some viruses these structures have been studied best so far, but they are expected to exist in humans alike. At lower temperatures, a stem-loop in the 5′-untranslated region of an mRNA might prevent the expression of a gene. When the temperature rises, however, that RNA secondary structure will melt away and the relevant gene can be translated. Human-pathogenic bacteria bring along such RNA thermometers. If it is cold outside, the little ones don't have to stretch themselves too much. But when the temperature approaches 37 °C or so, it is very likely that the bacteria have arrived in the body of a potential host. Time for the microscopic beasts to multiply and perhaps also to produce a toxin, or two.

6

Confirmatory How individual metabolic pathways are regulated—fine-tuned—probably is one of the most exciting areas in biochemistry. On a small scale, it is nearly always proteins or other biopolymers that are tickled, stimulated, or inhibited, by various compounds. On the larger scale of the whole organism, hormones—and finally the brain—come into play. It's fascinating to remember that a mere thought or even a feeling can switch on a biochemical metabolic pathway. Regulation is everywhere, even where it was not previously suspected, as we discover signal properties for more and more "standard" metabolites. A recent trend that will probably continue for a while. Cell-internal feedback loops are then enriched by cell-to-cell communication, and the sensing of changes in a cell's microenvironment.

▪ Here You Can Find More About Regulation

Franz Narberhaus from Bochum, Germany, has published extensively about those temperature-taking bacterial RNA thermometers—Twittenhoff et al. **2020**. An RNA thermometer dictates production of a secreted bacterial toxin. PLoS Pathog. 16(1):e1008184. ▸ https://doi.org/10.1371/journal.ppat.1008184.

Here is the concept that "boring" metabolites can become exciting regulatory molecules—Haas et al. **2016**. Intermediates of Metabolism: From Bystanders to Signalling Molecules. *Trends Biochem Sci*. 41(5): 460–471. Review. ▸ https://doi.org/10.1016/j.tibs.2016.02.003.

And for those who are not yet impressed enough, here are **three papers** that all have in common that they all highlight a certain metabolite, but all three highlight a different one!

ONE: Claudio Mauro is a fan of **lactate**—Certo et al. **2019**. Lactate: Fueling the fire starter. *Wiley Interdiscip Rev Syst Biol Med*. e1474. Review. ▸ https://doi.org/10.1002/wsbm.1474.

TWO: **Glc-6-P** is also something great—Rajas et al. **2019**. Glucose-6 Phosphate, A Central Hub for Liver Carbohydrate Metabolism. *Metabolites*. 9(12):282. Review. ▸ https://doi.org/10.3390/metabo9120282.

THREE: Ed Chouchani really likes **succinate**—Murphy & Chouchani. **2022**. Why succinate? Physiological regulation by a mitochondrial coenzyme Q sentinel. *Nat Chem Biol*. 18(5): 461–469. Review. ▸ https://doi.org/10.1038/s41589-022-01004-8.

6.7 How Could the Central Metabolism Have Originated?

As recently as 2000, the English biochemist Leslie Orgel equaled the emergence of the Krebs cycle solely to "magic". A magical cycle—somehow comparable to the magical roundabout in the English town of Swindon (▪ Fig. 6.4), a town not so

Fig. 6.4 **The magic roundabout in Swindon**. Extraordinary forms of traffic arrangements quickly earn themselves special names, be it in England, or elsewhere. For example, the giant roundabout in the southern English town of Swindon is called *Magic Roundabout*. This Magic Roundabout is composed of five smaller roundabouts and regularly causes desperate drivers and relationship crises behind the wheel. Probably not magically, but fueled by sulfate radicals, the Krebs cycle originally got started. Sulfate radicals for example, could come from the decay of persulfate salts. The spark that jumps across is represented by a sparkler. (Image credit: Traffic sign, Stephen, AdobeStock; Sparkler, sunakri, AdobeStock)

magical. Orgel's main argument was that the eight reactions of the Krebs cycle are only biochemically valuable when taken together—so the simultaneous random emergence of eight reaction centers in eight enzymes would have to be assumed. An event so unlikely, one could quickly resort to magic as an explanation.

Our understanding of how things might have come into existence, have changed in recent years, not so much for that special junction in Swindon, but more so with regards to the Krebs cycle. A fairly new theory on the origin of the Krebs cycle has been developed by biochemist Markus Ralser. His research group discovered that quite a few of the central metabolites have ALWAYS performed their respective reactions—even without enzymes and cells. We now know that ALL reactions of the Krebs cycle can occur spontaneously, when, just when, they are initiated by sulfate radicals.

That makes a lot of sense, doesn't it? The scenario could have been like this: A couple of reactions run spontaneously in the primordial soup. They form some sort of loop. Much later, enzymes are forming all around these reactions, to make them even more efficient. And then cell membranes close in around it. LUCA is formed; the frozen accident is readily trapped. This scenario also explains why the Krebs cycle is so strikingly different to glycolysis. In glycolysis, there are ONLY metabolites with a phosphate residue, at times even with two phosphates. In the Krebs cycle, NOT a single intermediate product is phosphorylated.

Catalyst The Krebs cycle is considered the central hub of metabolism. For a long time, it was unclear how it could have originated. Most recent research suggests that the reactions of this cycle could also run non-enzymatically under conditions as they were in the primordial soup. Non-enzymatic, pre-biotic reactions seem to be crucial for the emergence of life.

▪ For Further Reading

The word "magic" does not appear often in biochemical publications. Here you can find the original quote by Dr Orgel—Orgel. **2000**. Self-organizing biochemical cycles. *PNAS*. 97(23): 12503–7. ▶ https://doi.org/10.1073/pnas.220406697.

The rhetorical counter-question from Markus Ralser can be found here—Ralser. **2018**. An appeal to magic? The discovery of a non-enzymatic metabolism and its role in the origins of life. *Biochem J*. 475(16): 2577–2592. Review. ▶ https://doi.org/10.1042/BCJ20160866.

The author of this book also reported on one of the groundbreaking works from the house of Ralser—Mueller. **2017**. In the beginning, there was Sulfate. JournalClub. *BIOspektrum*. 23(3): 292.

Let There Be Light: The Only Genuinely Green Chapter

Contents

© The Author(s), under exclusive license to Springer-Verlag GmbH, DE, part of Springer Nature 2026
J. W. Mueller, *Ultimately Understanding Biochemistry*, https://doi.org/10.1007/978-3-662-71889-6_7

For most parts of this book, we have discussed the most popular model organism of biomedical research—the human body itself. Teaching the biochemistry of the human body is a biomedical necessity… And doing so comes in handy, because we don't have to deal with all the complicated biosynthesis pathways of individual amino acids and cofactors, and so on, and so forth. To understand where an essential amino acid or a vitamin is coming from, it is sufficient to label such compounds as "essential" and to remind ourselves to maintain a "good" diet, to take in sufficient quantities of this stuff. Job done. What remains are glycolysis, Krebs cycle, et al. And lets be honest, most of what we have learned about all of it applies to all organisms, anyway.

The metabolism of humans and their mammalian relatives comes with its very own restrictions. Plants, on the contrary, can convert fats into sugar with ease. Plants do so in the glyoxylate cycle, an anabolic variant of the Krebs cycle. In addition, we have already indicated that plants are world champions in many biosynthesis reactions, please be reminded of the chapter about isoprenoids (▶ Chapter 2). Making "green" isoprenoids, instead of "red" ones, tends to be much more diverse, even creative.

Talking about "red" and "green" color designations, do they speak to you? Specialists enjoy classifying the various areas of biotechnology into up to 10 different categories, all with their own color assigned! Here, we want to limit ourselves to the "red" biotechnology of animals and humans and the "green" one of plants. Please do not try to spoil even this streamlined color scheme, by coming up with green frogs or red houseplants. Thank you.

Everyone knows that **vitamin C** is healthy. It is so healthy that we, human beings, have unlearned how to make it, so we have to take it up from vitamin-C-rich diets, such as potatoes, or citrus fruits. For fans: Within the UK its highest amounts can probably be found in black currants (◘ Fig. 7.1). Other sources might be some effervescent, or bubbly, vitamin C tablets. Within the body, vitamin C is a redox-buffering substance. Not all mammals necessarily feed on potatoes, oranges, and black currants, however. So, how do these other animals get their vitamin C? Can they make vitamin C by themselves? A "simple" answer from a

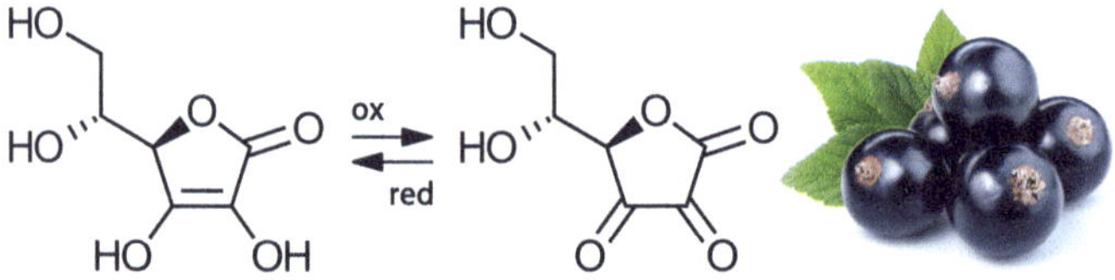

◘ **Fig. 7.1 Vitamin C protects us against oxidation, diseases and aging**. We usually get vitamin C from our diet in its reduced form (**left**). While absorbing it, vitamin C gets oxidized and converts into the form shown in the **middle**. Vitamin C is contained in every fresh food, more or less. Quite a lot is found in citrus fruits, but also in potatoes. Black current is regarded as the British record holder, at least in terms of vitamin C content. For fans, it was Norman Hayworth from Birmingham, UK, who figured out the molecular makeup of vitamin C. (Photo credit: Sea buckthorn, © emberiza ▶ stock.adobe.com)

genetic study from 2011 is: Both cat and mouse as well as pig and cow can **still** produce their vitamin C themselves. However, loss of vitamin C biosynthetic capacity has happened several times: True bony fish (but no other fish), bats (but no other close relatives), guinea pigs (but no other rodents) as well as all monkeys including humans (but with the exception of lemurs) have lost the ability to synthesize vitamin C. Plain and simple, right?

To illuminate this patchy genetic trait, at least in parts, the first dedicated gene of vitamin C biosynthesis, the gene for L-gulono-γ-lactone oxidase, broke down multiple times in different species....and because some of us (in this case, "us" are a random collection of animals) could cover our vitamin C needs through our diet (▪ Fig. 7.1), that genetic defect has been retained.

For *Homo sapiens*, that all worked well, until "mankind" came up with the idea of sailing the seas for many days to discover new worlds, ill-fed. In those times, vitamin C deficiency caused the disease scurvy. While modern-days sailors consume a diet providing sufficient amounts of this vitamin, the deficiency in vitamin C remains to be a problem in some parts of the globe.

Anything you can do I can do better; I can do anything better than you. This line of music from the Broadway musical Annie Get Your Gun can be applied to our plants, along with algae and some bacteria, in that they are in fact masters in quite extraordinary biosynthesis pathways. First, plants absorb water from the ground, or take it up from the morning dew… then they take carbon dioxide from the atmosphere, and convert this into sugar and oxygen, whenever sunlight is available—seemingly living on love alone.

$$6CO_2 + 6H_2O \rightarrow C_6H_{12}O_6 + 6O_2$$

After all the biochemistry this book has delivered so far, this net reaction of photosynthesis hopefully doesn't look that shocking anymore. But is it smashingly insightful instead? This reaction encompasses several processes that occur far from voluntary. Actually, none of the whole reaction would run on their own; a lot of biochemical "convincing" is needed. What follows is a brief overview of plant photosynthesis, illuminating exactly those steps where biochemical and enzymatic "help" is delivered, to help plants to drive this very reaction to completion.

Tenants can be annoying. Plant cells accommodate chloroplasts as sub-cellular tenants, in addition to mitochondria, all stemming from endosymbiosis events—where one cell engulfs another cell. As a consequence, plant cells have to deal with two types of partially autonomous cell organelles with their own genomes, their own protein expression machinery as well as countless import events into these organelles on the one side and retrograde signal transmission out of these organelles on the other side. Any gene expressed by these organelles and any other metabolic behaviors must be tightly coordinated with the nuclear genome, somehow. Chloroplasts are known for photosynthesis, but—similar to mitochondria—they play an important role in a good number of other biosynthesis pathways, too.

Cooperative Just take a look around you. Almost everywhere in the world you can find plants or their more compact relatives—algae or photosynthetic bacteria. Plants carry out photosynthesis in special organelles, the chloroplasts. If there is some supply of water and light, plants may be able to survive even in the saddest and darkest places, such as some office plants. No, I don't show you a picture of my office here.

Further Reading

Located in Bergen, Norway, Iain Johnston thinks about our organelles—Johnston IG. **2019**. Tension and Resolution: Dynamic, Evolving Populations of Organelle Genomes within Plant Cells. *Mol Plant*. 12(6): 764–783. Review. ▶ https://doi.org/10.1016/j.molp.2018.11.002.

If you've got it, flaunt it! Vitamin C synthesis in various vertebrates—Drouin et al. **2011**. The genetics of vitamin C loss in vertebrates. *Curr Genomics*. 12(5): 371–8. ▶ https://doi.org/10.2174/138920211796429736.

Exploring different worlds… About red and green metabolic pathways—Günal et al. **2019**. Sulfation pathways from red to green. *J Biol Chem*. 294(33):12293–12312. Review. ▶ https://doi.org/10.1074/jbc.REV119.007422.

7.1 Pump it, Pump it: Sunlight Photons Power Electron Pumps

A major part of the gigantic photosynthesis apparatus consists of two huge protein complexes, the photosynthesis complexes I and II. Photosynthesis complex I is a round floating disc, swimming in the membrane; it is densely stuffed with colored cofactors. These cofactor antennae (◘ Fig. 7.2) act like solar panels and absorb blue and red light; green light is left over and confers the beautiful green tone to most plants. Having quite many antenna pigments comes handy, as the energy can jump back and forth between different pigments, almost without any loss due to near-optimal Förster resonance energy transfer. This setup really does a great job. It massively increases the efficiency of the entire light-harvesting exercise: catching photons and getting an electron excited about it.

The activated electrons eventually reach the central electron transport chain, which consists of a few chlorophyll molecules, some phylloquinone cofactors, and three iron-sulfur clusters. Driven by the energy of light, this machinery pumps electrons from plastocyanin in the thylakoid lumen to ferredoxin on the stroma side of the chloroplast. The whole thing is distantly similar to separating eggs when baking. As more and more electrons are pumped to only one side of the membrane, a proton gradient builds up simultaneously. When this proton gradient is allowed to flow back through the inner mitochondrial membrane, it will ultimately power ATP production by spinning ATP synthase around.

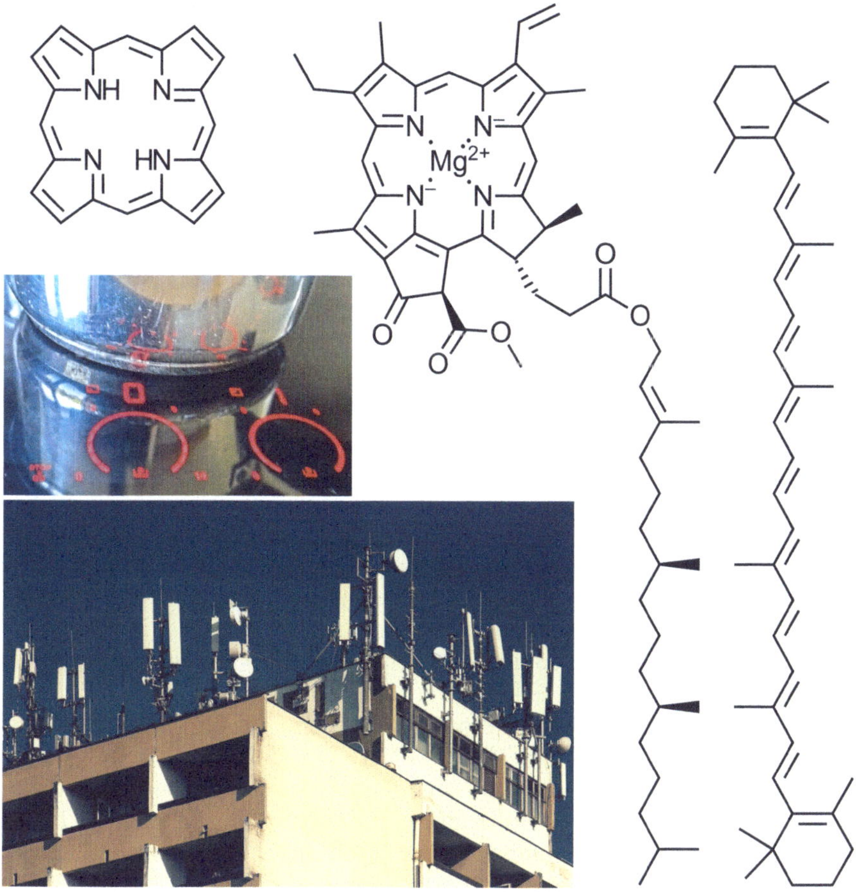

Fig. 7.2 **Antenna complexes in the green and on the roof**. The crown-like molecule porphyrin is great at tightly holding metals. Adding a magnesium ion and a few notable extensions here and there, results in the structure of chlorophyll (chlorophyll A is shown in the middle). And then we have lycopene on the **right**, which belongs to the group of carotenoids. It is abundant in tomatoes, if you want to get some lycopene easily. Several types of antenna pigments, understand each other "blindly". They can exchange energy between each other without radiation—somewhat similar to your induction stove and a suitable pot. Light-harvesting complexes are crammed with many other molecules that look similar in size and function, making these big complexes most efficient absorbers of energy from the sun. (Image reference: Antenna forest, © Zsolt Biczó ► stock.adobe.com)

7.2 Let There Be Air, Once We Get Hold of Water's Beloved Electrons

We know that hydrogen and oxygen are really fond of each other. The reaction between these two elements is also known as the *knallgas* reaction—there's a real bang behind it! Water is created in this way. This time it's not about the special properties of water, the very topic with which this book began (▶ Chapter 1). Instead, photosynthesis is about how to pick water apart—how to best destroy it. Anyone who knows oxygen might also know that this element has the second highest electronegativity, in the periodic table, after fluorine. Oxygen therefore does not "willingly" give away its electrons, earned in previous redox reactions.

In theory, it is quite simple to destroy water. All you need is a burning magnesium chip, dipped into water. Magnesium burns at more than 2500 °C. That's enough thermal energy to split water into its components—**pyrolysis** is the magic word here. The oxygen thus released promotes further combustion of the metal.
7 Remarkable, isn't it. Plants can destroy water at room temperature, however; and they can do it with ease.

The ultimate water destroyer, the absolute specialist for water destruction, is photosynthesis complex II. It is slightly more egg-shaped than photosynthesis complex I. Complex II has fewer antenna pigments directly bound, instead it interacts with antenna protein complexes, richly stuffed with antenna pigments. **Photosynthesis complex II** has a bizarre arrangement of four manganese, one calcium, and five oxygen atoms in its center. This structure is also referred to as the oxygen-producing center (◘ Fig. 7.3). This Mn_4CaO_5 cluster has strongly delocalized—almost freely floating—electrons, the cluster functions like a kind of battery.

As soon as light shines on this complex, the battery does not charge, but discharges. It is systematically pumped empty of electrons. Then two molecules of

◘ **Fig. 7.3** **A magnesium torch and the oxygen-generating center**. A burning sparkler serves to remind us of the beautiful, bright glow of a burning magnesium chip, even though the sparkler most likely will not contain a lot of magnesium. Next to it is the oxygen-evolving center (OEC) to be seen—an awkwardly skewed cube of manganese (purple), calcium (cyan) and oxygen atoms (red). At a distance, the rest of a tyrosine is depicted. Somewhere between the OEC and the tyrosine, there are two water molecules bound. Please refer to the main text to learn what is going to happen to them. (Image credit: Sparkler, © sunakri ▶ stock.adobe.com; OEC, own structure visualization after ▶ https://doi.org/10.2210/pdb1S5L/pdb)

water come along—and—BANG! In one single go, the electron-hungry manganese-calcium cluster grabs four electrons from the two water particles. The two water molecules are literally torn apart immediately. A proper molecule of elemental oxygen forms, which diffuses away. The remnants are four protons, that diffuse off, on their own.

Complex What we call "light reaction", is where oxygen is produced; in addition, plenty of ATP and NADPH is generated. But so far, not a single atom of carbon has been caught from air and chemically fixed.

■ **For Further Learning**

By now, a lot is known about the structure of the various components of the photosynthesis apparatus. I recommend starting with the tutorials of the Protein Data Bank, for example with this one ▶ https://pdb101.rcsb.org/motm/244 [Accessed 2.7.2022].

7.3 Probably the Slowest Enzyme on Earth Pulls Carbon onto the Dark Side of Photosynthesis

It's amazing what the enzyme RuBisCO does. In the dark, RuBisCO binds a happily free-floating carbon dioxide molecule, forces it into the conformation of a carboxyl group, and connects it to the recipient molecule ribulose-1,5-bisphosphate. Some of us are now thinking of some lords and rings. Something like this, maybe: *One acceptor to rule them all, one acceptor to find them, one acceptor to bring them all, and in the darkness bind them*... For others, ribulose-1,5-bisphosphate simply fixes some carbon.

In ▶ Sect. 5.5, we had mentioned already the carbon dioxide-fixing enzyme RuBisCO. The enzyme is far from being perfect. It is quite slow and manages just three carbon dioxide fixations per second Moreover, RuBisCO is also negligently sloppy. The enzyme quite often mixes up its actual substrate CO_2 with molecular oxygen—both are quite similar in size. An oxygenated sugar is the unintended consequence, which costs the plant cell quite a bit of repair effort. Nevertheless, almost all living beings on Earth depend directly or indirectly on this one reaction. Plants compensate for the extremely poor catalysis rate of RuBisCO by simply producing tons of this enzyme. The more, the better This makes RuBisCO the protein that probably exists most on our planet (**napkin XIV**).

Napkin XIV: How Much RuBisCO Is There on Earth?

A lot of carbon is bound on Earth, ≈ 100 Gt/year = **10^{17} g carbon/year**

This is the net fixation rate. Because of plant respiration, the gross to net factor is ≈ 2

The molecular weight of carbon is 12 g/mol

RuBisCO has a molecular weight (per subunit) of 70,000 g/mol

A year consists of $60 \times 60 \times 24 \times 365$ seconds $\approx 3 \times 10^7$ seconds

RuBisCO manages a maximum of 2–10 carbon atoms per second, but on average, RuBisCO fixes ≈ 1 carbon atom per second

All of the above together makes 3.9×10^{13} g ≈ **4 x 10^{10} kg active RuBisCO**

This mass expressed in number of cars (1.4 t average weight) ≈ 28 million cars

World population 8.2 billion people = 8.2×10^9

Total mass of active RuBisCO per earthling ≈ **5 kg/person on Earth**

For fact check: I have found this calculation on a napkin in this paper by Ron Milo—Phillips & Milo. **2009**. A feeling for the numbers in biology. *PNAS*. 106(51): 21465–71. ► https://doi.org/10.1073/pnas.0907732106.

7

Roughly speaking, RuBisCO does the following: To warm up, a specific lysine side chain grabs a CO_2 molecule, to become a carbamate. This carbon dioxide molecule will not end up in the sugar, it is purely catalytic—a kind of on and off switch. With the help of a few other amino acids, this carbamate then complexes a magnesium ion. Then RuBisCO binds the CO_2 acceptor molecule ribulose-1,5-bisphosphate and converts it from its keto form to its enol form, a form of activation. The next CO_2 molecule that comes along, first docks to the magnesium, and then binds to ribulose-1,5-bisphosphate. The product of this reaction is unstable and breaks down immediately into two identical units of glyceraldehyde-3-phosphate. Carbon fixation completed.

Next up is the Calvin cycle that features reactions very similar to the ones from the pentose phosphate pathway (► Chapter 6). Sugars with three, four, five, six, and even seven carbon atoms are fused and cleaved again—being pushed back and forth at near-equilibrium conditions—until all the two triple sugars that RuBisCO provided have become decent C6 glucose sugar or at least halfway decent C5 ribose. All this is catalyzed by just two enzymes—transaldolase and transketolase—following the knitting pattern *two-to-the-left, three-to-the-right*.

▪ Suggestions for Further Reading

A comparison of different RuBisCO enzymes and their lousy catalysis rates—Sage. **2002**. Variation in the k_{cat} of Rubisco in C3 and C4 plants and some implications for photosynthetic performance at high and low temperature. *J Exp Bot*. 53(369): 609–20. ► https://doi.org/10.1093/jexbot/53.369.609.

A pretty clever approach to finding faster RuBisCOs—just take a closer look at what Nature has to offer—Davidi et al. **2020**. Highly active rubiscos discovered by systematic interrogation of natural sequence diversity. *EMBO J*. 39(18): e104081. ► https://doi.org/10.15252/embj.2019104081.

7.4 Plants: Biosynthesis World Champions of Secondary Metabolites… and… WOOD

Plants can move. You might think of daily movements of your potted plants at home, following the sun, or the astonishing spreading of seeds for plant reproduction. Overall, the movement of plants is limited, however. Plants cannot simply run away from predators. That's probably why plants have become masters of secondary metabolites—all those bitter substances, spices, and toxins, were actually to spoil the appetite of caterpillars, gall wasps, and other voracious creatures. Somewhat ironic that it is us humans, who have learned to love some of these defense mechanisms to flavor our diet.

A special class of plant secondary metabolites are **glucosinolates**. They occur in almost all cruciferous plants—for example in various types of cabbage, in mustard, and in horseradish. The general structure of glucosinolates is shown in ◘ Fig. 7.4. At the center of all glucosinolates is a carbon atom, provided by an organic residue, an amino acid, in many cases. That central C-atom is bound to a sulfur atom, which is again bound to a glucose unit. From our carbon center, a third bond, a double bond, goes to a nitrogen atom, which is in turn linked with an OH group. This structure is an aldoxime group. Doesn't that sound pretty? Finally, that OH group is esterified with a sulfate residue. Done. The central S-C(R)=N(O) group is typical for all glucosinolates.

Chemically well-crafted, these **sulfur-containing** aldoxime-**compounds** resemble small stink bombs. When the plant is bitten or otherwise damaged, glucosinolate compounds spontaneously decompose and mustard oils are formed, which can taste bitter or spicy, they might even be poisonous. Well, what do we humans do with glucosinolates? We make all sorts of mustard, for a start. As glucosinolates give bitter notes to some vegetables, plant breeding tried to reduce glucosinolate contents here or there. Maybe a thing of the past, as several glucosinolates are now attributed health-promoting properties. Are we back to healthy mustard again?

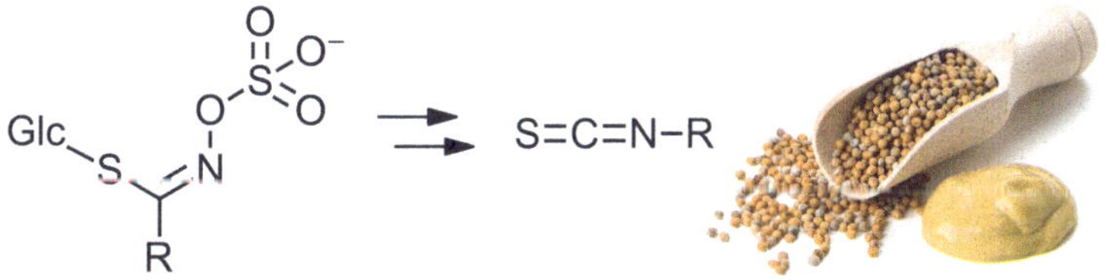

◘ **Fig. 7.4 Glucosinolates for defense and for delicious mustard**. The general structure of glucosinolates: Around the central S-C=N(O) unit, a sugar (usually glucose), a sulfate group, and various organic residues are arranged. It is this organic residue where glucosinolates show the most diversity in. It doesn't take much for glucosinolates to fall apart: If glucose is cleaved off, sulfate is released and isothiocyanates are formed through internal chemical rearrangements. Hopefully, these mustard oils protect against predators. They may also be good for our health. In any case, isothiocyanates produce a delicious aroma in mustard or as an aftertaste in various types of cabbage. (Image credit: Mustard seed and mustard, © Diana Taliun ► stock.adobe.com)

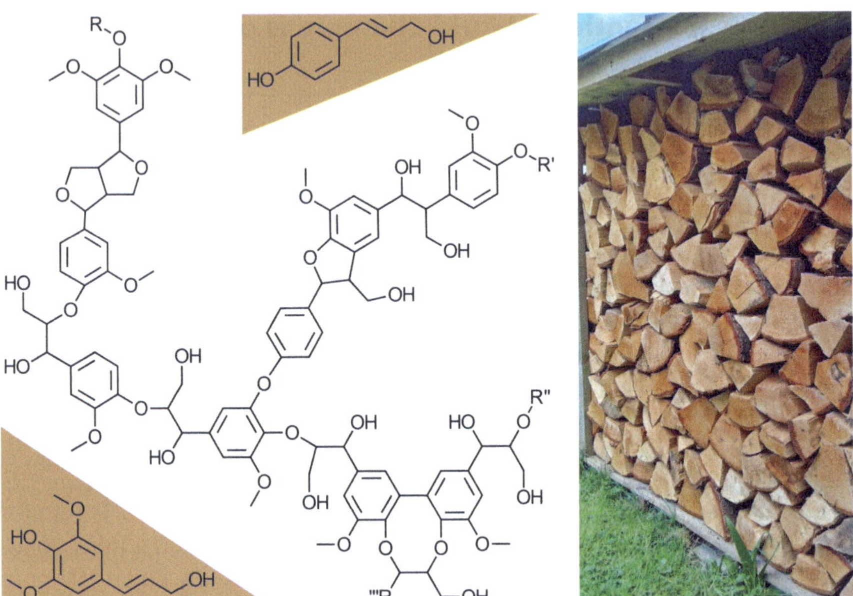

7

Fig. 7.5 **Lignin & Wood**. You can see monomeric building blocks of lignin at the bottom left and top right. Activated radically, these building blocks chain up and down, in some uncontrolled polymerization. What a mess. All sorts of binding types can be found in lignin. Ether bonds seem to be the most common, but there are also C-C bonds between alkanes and/or aromatics as well as examples of creative cyclization. The strategic advantage of this jumble is that a single enzyme, such as cellulase, could not do much degradation here. Chemically, not much happens either with lignin, when cooked in acid or in alkaline. By the way, lignin not only internally cross-connects, it also cross-links with other biopolymers and therefore protects other components of the cell wall, especially cellulose, from degradation by microorganisms. In the long run, however, fungi and bacteria will win, always, but only because they deploy a plethora of dedicated enzyme machinery. Fun fact: the term lignin is derived from the Latin *lignum* which means wood, so the title of this figure actually is "wood & wood". (Image credit: JWM, 2022)

Having discovered exciting facets of glucosinolates, we could talk about so, so many plant-based secondary substances—just think of caffeine, morphine, digitoxin, tetrahydrocannabinol, nicotine or perhaps of natural fibers such as cotton or hemp. All of them are certainly worth several paragraphs, or even pages, but please read about them somewhere else. I conclude this chapter differently. When a plant is damaged, the damaged area should be sealed as quickly as possible. For this purpose, plants have developed a very special polymer glue. This or something similar could be the beginning of the evolutionary success story of a rather special secondary metabolite. We are talking about wood, well, first of all about **lignin** (Fig. 7.5).

Plants colonized the land of our planet about 450 million years ago. These first settlers were vascular plants that relied on lignin. Probably first created as a protection against grazing, lignin soon served to reinforce the water pipes within the plant—the xylem. Only this fortification allowed water to be transported from the root to the leaves over longer and longer distances. Supported by lignin and cellulose, those longer distances soon became even longer as trees stretched their crowns into the sky. **Wood** has served as an important building material for construction, ever since. As burning fuel it provided warmth, and it became the starting material for important things such as paper or charcoal; think of your BBQ. We'll leave it here, for now.

Further Reading

A little more about the terrestrialization of plants, lignification and possible applications –Renault et al. **2019**. Harnessing lignin evolution for biotechnological applications. *Curr Opin Biotechnol.* 56: 105–111. Review. ▶ https://doi.org/10.1016/j.copbio.2018.10.011.

7.5 Healthy and Productive Plants in Times of Climate Change

Great challenges are ahead of us, not only because of the now gone Corona pandemic. It is necessary to supply an ever-growing world population with high-quality and sustainably produced food—a mammoth task. At the same time, valuable farming areas are being lost due to flooding, erosion, construction, and climate change. What we should do about climate change, exactly, is being heatedly debated about—particularly in Central Europe, sometimes at the edge of missing the point. Let's discuss one thing after the other, in this chapter.

Livestream from the inside of a plant cell: RuBisCO is trying again to distinguish between CO_2 and O_2 molecules. At higher temperatures, especially, RuBisCO is quite overwhelmed with this task. That is not good news. The increasing rate of photorespiration reduces the yield of photosynthesis, on more and more warmer days. All this only comes on top of water shortage, as water usually becomes scarce with rising temperatures. Plants close their stomata then. Evaporation decreases. But carbon dioxide no longer enters the leaf either.

Nature solved this problem by "inventing" C4 plants. They are called C4 plants, not because of a new model of a French car manufacturers, but because of a different type of carbon fixation. Unlike outlined in ▶ Sect. 7.3, the first product of carbon fixation has four carbon atoms (C4), instead of three (C3). This is what these plants are named for—C4 plants produce some four-carbon compound and that happens spatially separated from the actual dark reaction. That additional reaction represents a form of pre-fixation of carbon dioxide; thus, CO_2 can be enriched near RuBisCO. Been "spoon-fed" with CO_2, RuBisCO can work significantly more efficiently, while also reducing photorespiration, on warm days and in times of water scarcity. Corn and sugarcane are great examples of agriculturally significant C4 plants.

How cool is this C4 metabolic pathway. When it comes to the question about how it might have originated, things get messy rather quickly (somewhat similar to the vitamin C story in ▶ Sect. 7.1). The C4 mechanism has probably evolved independently multiple times, in various plant families. Probably proof that it is a clever solution to an urgent problem. Experts speak of one of the most complex polyphyletic traits that we know.

A couple of thousands of years ago, we humans were just learning to cultivate some wild plant and grass species, to harvest them, and to somehow make edible meals out of them, maybe even tasty ones. Unfortunately, back then, we did not know yet about the C4 metabolism of some plants, but not of other plants, nor about our own morbid passion to change the global climate—a passion that creates suffering. A bit of bad luck—almost all of our crops are of the old C3 plant type. Making these C3 plants more climate-resistant takes a lot of effort. "Just" breeding existing wild C4 plants for yield sounds utopian. Couldn't we give our existing C3 crops a boost? Why not try to somehow turn potato or wheat into a C4 plant? A great idea. Moreover, an idea that many breeding researchers and plant physiologists have already been working on intensely.

Alternatively, could we perhaps get RuBisCO a little bit more on track? This project even sounds easier. From many, many plant species, the protein sequences for their individual RuBisCO have been collected. We also know the behavior, the enzymatic activity, and some other parameters, of some of these RuBisCO enzymes. Based on this, we should be able to develop a super-charged RuBisCO. Then it would "only" be a matter of replacing a single gene, the lazy, old enzyme with this great, new RuBisCO. Fingers crossed.

After all, plants are the foundation of our existence, regardless of how technologically advanced we become. With all what we discussed in this chapter, it is necessary to continue optimizing our crops using the most modern methods. Very importantly, this is NOT about interventions "done to Nature". It is about further optimizing useful plants, a process that our ancestors started in the Stone Age, already—plant breeding as always, just with newer tools available.

We have discussed plant optimization in light of higher yields and climate compatibility, so far. There is so much more, we could be aiming for: Plants could certainly be made more resistant to pests or more tolerable for humans and animals. The former by introducing resistance genes—this could be a molecular receptor, for example, that allows the plant to detect an infestation of a fungus more quickly so that it can launch its biochemical counterattack in a timely manner. The latter by removing a gene—wouldn't it be great to remove the peanut allergen from peanuts or to produce gluten-poor wheat? Why should we put aside all our scientific knowledge and all these promising approaches, only to instead romanticize "the past", "that" time where everything was better. You pick it, what "the past" actually mean—perhaps the nineteenth century? Or even the Stone Age?

And then there is the genuine biotechnological approach. One could simply take simpler plants—for example, single-celled algae—and cultivate them on a large scale, to produce all kinds of delicious and high-quality products from them; one wouldn't even need valuable arable land. Why not draw inspiration from plants and recreate their organizational and functional principles, starting with an enhanced RuBisCO? This brings us to the realm of synthetic biology. Perhaps we could replicate photosynthesis in an artificial cell, or in cell-free systems. Some things in this chapter could become reality relatively soon; a few others still sound like dreams of a likely, but distant future.

Cultivate Plants are super! They not only master photosynthesis; they can also produce many great substances from their secondary metabolism. We have gone into a little more detail on glucosinolates and lignin. We need to take care of our plants—this includes protecting our environment and climate, as well as handling our crops, professionally.

Navigation

Already in ▶ Sect. 2.5, we discussed plant secondary substances from isoprenoids.

It is him again. Ron Milo knows all about RuBisCo—de Pins et al. **2025**. Rubisco is slow across the tree of life. *Proc Natl Acad Sci U S A*. 122(47):e2501433122. ▶ https://doi.oprg/10.1073/pnas.2501433122.

Oxygen comes at a cost—Hamilton. **2019**. The trouble with oxygen: The ecophysiology of extant phototrophs and implications for the evolution of oxygenic photosynthesis. *Free Radic Biol Med*. 140: 233–249. Review. ▶ https://doi.org/10.1016/j.freeradbiomed.2019.05.003.

Current insights into the weird "phylogeny" of C4-genes, okay, a bit "confusing"—Casola & Li. Beyond RuBisCO: convergent molecular evolution of multiple chloroplast genes in C4 plants. *PeerJ*, **2022**; 10: e12791. ▶ https://doi.org/10.7717/peerj.12791.

Quite a sensational message: A team from Marburg, Germany, succeeded to recreate CO_2-Fixation in artificial cells—Miller et al. **2020**. Light-powered CO_2 fixation in a chloroplast mimic with natural and synthetic parts. *Science*. 368(6491): 649–654. ▶ https://doi.org/10.1126/science.aaz6802.

What Evolution Really Is About...

Contents

© The Author(s), under exclusive license to Springer-Verlag GmbH, DE, part of Springer Nature 2026
J. W. Mueller, *Ultimately Understanding Biochemistry*, https://doi.org/10.1007/978-3-662-71889-6_8

Everybody is constantly talking about evolution. Some use the word in the sense of "personal development". Some think of "strategic decisions". Others plainly mean (hopefully good) "progress" or even some sort of "improvement". All of these meanings interpret quite generously what we understand by biological evolution in a sociological sense. "We", at this occasion, are Charles Darwin from the English Midlands and his intellectual heirs (◘ Fig. 8.1).

Darwin published his famous book "On the Origin of Species", in 1859. What exactly did Darwin do there? You are invited to pick up a copy of his book, to touch it with your own hands, maybe in a library close to you… I personally ordered a 1961 copy of the book up from our research reserve. Then just browse through it. You would see a long list of examples of evolution, a very long list indeed... Implying something like: Here you see it, and here you see it, and here, and here, and here too…. and here as well…

The persuading power of Darwin's book is not only the book's content, but also the sheer printed product. A reprint of the first edition has 540 pages. Some might say: The book is very much accurate, especially when thrown at close range… All of this sounds less scientifically revolutionary than we sometimes consider Darwin to be; it more documents hard work collecting and tabulating the evidence. Perhaps the time simply had come for the thoughts that Darwin stubbornly stuffed into his book. Most of the arguments presented had been there before anyway, in one form or another. His ideas greatly influenced other sciences; they impacted on culture and society in general. Foremost however, Darwin's ideas revolutionized all aspects of biology. Some say Darwin made biology a real science. Darwin started a new era.

◘ **Fig. 8.1** **Darwin is present everywhere in the English Midlands.** People in England are quite proud of Darwin, of course. … but being British, people don't show their affection too openly. Darwin is omnipresent in his birthplace Shrewsbury near Birmingham. You can see him also in Birmingham itself, for example, at the University of Birmingham, where he is one of nine wise men looking down on our students—in the row of scholars, he is the first from the right. For Darwin's 200th birthday, a beer was launched in his honor. Having a pint of Darwin's Origin is excellent for pondering and discussing about aspects of the theory of evolution. (Photo credit: Photos of UoB, Shirley Knauer, 2014, reproduction with kind permission; Darwin's Origin, JWM, 2014)

In 1973, Theodosius Dobzhansky said so aptly:

> *"Nothing in biology makes sense, except in the light of evolution."*—Dobzhansky. **1973**. Nothing in Biology Makes Sense Except in the Light of Evolution. *American Biology Teacher*. 35(3):125–129. ▶ https://doi.org/10.2307/4444260.

In addition to his books, one can approach Darwin by looking at his very interesting way of life. One could discuss controversies about Darwin's scientific actions. Or, one could illuminate his problems with mental health. We know quite a bit about this topic from his many letters, one containing this quote:

> *"But I am very poorly today & very stupid & hate everybody & everything."*—Charles Darwin, in a letter to Charles Lyell, 1861.

Back in the days, the **core thoughts of Darwin's** probably were that living beings evolve at all, that evolution actually takes place—instead of species being eternal entities. A logical consequence is phylogenetic kinship—a pedigree tree of all living beings. The main creative force here is **natural selection** that Darwin put into the proverb "*Survival of the Fittest*".

The tricky point, which was fiercely argued about during Darwin's lifetime, and way beyond, was how this "selection" was supposed to work, exactly. The first point of contention was that selection **operates without direction**. There is no greater plan-to-implement... With absolute certainty, we humans are NOT the crown of creation. Instead, some concerned citizen of planet Earth might mumble something sounding like "rather we are the downfall of the planet". The other point of contention was and still is, at which **organizational level** natural selection operates, something that we will discuss in the next paragraph.

- **Here are two overview articles on the general topic of evolution**

Pretty cool paper, asking how to find an evolutionarily suitable model organism for the respective research question—Bolker JA. **2019**. Selection of Models: Evolution and the Choice of Species for Translational Research. *Brain Behav Evol*. 93(2–3): 82–91. Review. ▶ https://doi.org/10.1159/000500317.

Somewhat theoretical, but very comprehensive—Papale et al. **2020**. Networks Consolidate the Core Concepts of Evolution by Natural Selection. *Trends Microbiol*. 28(4): 254–265. Review. ▶ https://doi.org/10.1016/j.tim.2019.11.006.

8.1 Survival of the Fittest

So, at which level does selection work? Let's start with **population against population**—one species "displaces" another species in its biological niche. On this topic, I have to think of the poor mallards which are not doing so well in many European city parks, as they are increasingly being displaced by Canadian wild geese and other birds. Similar scenarios occur again and again when newly introduced species, geekily named neobiota, become invasive plagues and displace native species.

One of the first global plagues of the modern era is the Norwegian brown rat *Rattus norvegicus*. It only spread from Europe, not the opposite. So cool, sarcastically speaking, that it is "us", the Europeans this time, who set a trend.

Another example of population vs population could have been *Homo sapiens* against *Homo neanderthalensis*, as they opposed each other on the banks of the Düssel river in the shallow Neander Valley, not too far away from what would become Düsseldorf later. Through the analysis of ancient DNA, at the latest, we now know that the present human (*sapiens*) has displaced the Neanderthal from the Neander Valley and from many other regions, globally. While withdrawing from the evolutionary battlefield, Neanderthals strangely left remnants of their genomic DNA behind in our genome... Geneticists surely can explain to us how that might have happened during this conflict. One thing is certain, we all are related to the Neanderthal, to varying degrees. Some of us harbor more Neanderthal DNA, others less of it. Somehow that had been suspected already; maybe on a Saturday evening, when walking through the historical old town of Düsseldorf, which for its many pubs and bars, is also known as "the longest bar in the world".

We all know the version of Darwin's thought in which the one who survives is the strongest, we here interpret the "fittest" in *survival of the fittest* as physical strength. When we think of the zebra in the extended African grasslands and the lion in the shadow of a tree, this thinking suggests that the lion is contenting the zebra for survival. This thinking is wrong. ◘ Figure 8.2 probably better illustrates who really fights with whom for survival, in the sense of Darwin.

It is the individuals within the dazzle of zebras, who are fighting for the passing on of their genes, here we are dealing with an evolutionary process **individual**

◘ **Fig. 8.2 Two scientists in the sands of the Sahara.** In the desert, two scientists suddenly face a lion. One of them says: "We have to run and we have to do it right now!!!" The other: "That's pointless. We can't out-run a lion." The first one again: "Well, my dear colleague, actually I don't need to out-run the lion. I just need to out-run... well... you..." Conclusion: Not always is it immediately obvious, who is fighting with whom for survival. (Photo: JWM, 2022)

against individual within a population. The evolutionary biologist Richard Dawkins described natural selection as the **non-random survival** of **random variants**. Those two zebras represent random variants of a population, and the lion then becomes the embodiment of natural selection. The zebra that runs faster, that has funnier striping patters making it harder for the lion to focus on, or that features any other selection advantage, will survive.

Richard Dawkins took it even further and defined not the individual organism or perhaps a certain population of organisms as **the unit of selection**, but the **genes** contained therein. Accordingly, every gene variant "wants to spread" more than other variants thereof. It is the genes in the one and the other zebra that are competing with each other. Are all of us mere servants to our genes?

Let's think of the Corona pandemic between 2020 and 2023. We all were in the middle of a huge evolutionary endeavor of viral genes. So many variants of the virus constantly emerge through spontaneous mutations. Most of these variants are not relevant, we might see them shortly by sequencing, and then they are gone again. Some variants, however, we will see more often than we probably are comfortable with—in the past they had names such as Delta, or Omicron. Fun fact: The original CoVID-19 genome sequenced in Wuhan in late 2019 is no longer circulating in the human population. It has completely been replaced by newer variants. The whole pandemic showed us the emergence of variants and natural selection in its purest form. Perfectly cool and scientifically enlightening, but hopefully we won't be in another pandemic too soon again.

By the way, **evolutionary anthropology** is the study of what made us human, evolutionary. This discipline has compiled a multitude of molecular events that have occurred during the origin of humans—whether they have contributed causally to shape *Homo sapiens* or not. Quite a few changes have affected the brain, not too surprisingly. The course of childhood and puberty was radically remodeled—humans remain children much longer than most mammals. There is the old debate whether our sense of smell has decreased in the process of becoming human or not. Yes, we have many smell receptors that have become inactive "pseudogenes"; but was that really a necessary event for starting to live together in small caves, tents, or huts? And there are "old" diseases like the plague, that could have made us somewhat more resistant to new diseases, such as HIV. The topic was briefly brought up already in ▶ Chapter 1. I also want to mention two tough nuts for evolutionary theorists—the exact explanation of how beauty (let's think of the peacock's wheel) and homosexuality can be maintained evolutionarily.

Within human evolution, **cooperation** is the key term. Isn't it a comfortable thought that "the fittest" in Darwin's theory might be the best cooperator? Together we are strong! Modern genomic research shows us more and more the exact timeline, when we, *Homo sapiens*, came to live with ox and donkey. Of all the domestic animals, the interaction between humans and dogs seems to be of foremost interest (see **meandering text box**).

Meandering: How Homo sapiens Came to the Dog

Cat, pig, and cattle were domesticated by sedentary human groups who had begun to undertake agriculture and animal husbandry. The dog is the only pet that was domesticated by hunters and gatherers even before that time. Although time estimates in DNA-based archaeology are somewhat difficult, there is increasing evidence that the dog is the oldest companion of modern humans. The alignment between dog and human is now dated to have happened between 40,000 and 15,000 years ago.

But then the scientific debate begins: Did the dog domesticate itself by getting used to scavenging the meat scraps left behind by humans? Or was it humans who took in orphaned wolf pups and lovingly raised them? Regardless of how and when exactly it happened, for tens of thousands of years, humans and dogs have been roaming the lands together. It can be assumed that this cooperation gave *Homo sapiens* another advantage in the displacement competition with other hominids. Here, Darwin's phrase "*survival of the fittest*" could be interpreted as "survival of the **best team**".

■ **For further reading**

When humans spread in the Young Stone Age in Europe, they already had their "best friend" by their side—Ollivier et al. **2018**. Dogs accompanied humans during the Neolithic expansion into Europe. *Biol. Lett.* 14(10): 20180286. ▶ https://doi.org/10.1098/rsbl.2018.0286.

A diet, energy, and nutrition calculation. In Ice Age winters, our ancestors probably had more hunted meat than they could use themselves. So, they could share resources—Lahtinen et al. **2021**. Excess protein enabled dog domestication during severe Ice Age winters. *Sci Rep.* 11(1):7. ▶ https://doi.org/10.1038/s41598-020-78214-4.

A significant difference between dogs and wolves are their facial muscles; they are much more mobile in dogs. Maybe it was those dogs that could look lovably sweet (or "loyally stupid") that were preferred in breeding—Kaminski et al. **2019**. Evolution of facial muscle anatomy in dogs. *PNAS.* 116(29): 14677–14681. ▶ https://doi.org/10.1073/pnas.1820653116.

Creative Evolution is the non-random survival of randomly generated genetic variants. It occurs at all levels of biology, between genes, between individuals, and even between populations or whole species. In the next section, we will briefly discuss how genetic variance arises, for example when transposons, self-splicing introns, and retroviruses have their "say". More about that topic however in established genetics textbooks.

8.2 Lamarckism and Darwinism Do Not Have to Clash with Each Other

Jean-Baptiste Lamarck certainly was a clever man. He is considered the founding father of modern zoology of invertebrates, he advanced systematics in biology, and he did many other good things. But above all, he is known for being hopelessly wrong when it comes to evolution. History portrays him as a poor fool, ridiculed and laughed at.

Supporters of Lamarck engaged in a historical dispute with "Darwinists"—about how variance could arise and then be inherited. The core theme of Darwinism, succinctly formulated by Richard Dawkins—evolution is "the non-random survival of random variants"—seemed to be absolutely incompatible with Lamarck's views. He postulated the heritability of changes caused by the environment.

There is nothing wrong with these views, as of now. The weak point of Lamarck's views was that these changes could allegedly be "intended" or "directed". Something along these lines: the giraffe "wanted" to reach the higher branches, and did so again and again, and therefore got the long neck (and passed it on to its offspring). It is this purposefulness in Lamarck's views, which Darwin had expressly driven out of biology. This "purpose" opened a back door for all sorts of dubious pseudo-scientists, which was not exactly beneficial to Lamarck's reputation.

From today's perspective, the big dispute back then was a lot of hot air—a dispute that was mainly about beliefs and sentiments, less so about facts. The actual "bone of contention", the material of inheritance and its mechanisms were far from being discovered. Neither DNA in general, nor the concept of genes, nor the actual nuclear bases in particular were known. So, who was right, Darwin or Lamarck? The answer to this question might be somewhat milder, today.

Nature uses both ways: the strict DNA-dependent inheritance from generation to generation **AND** the epigenetic modification of chromatin. We have already mentioned changes to those curler-histone proteins—assumed to be short-term and initially non-heritable changes in genetic material or to the nucleosomes in its vicinity (discussed in ▶ Chapter 3). The whole thing becomes more complicated because some of these **epigenetic markings** themselves are quite permanent, as they influence the methylation of DNA—at cytosine bases, to be precise. Then the spontaneous deamination of cytosine can no longer be repaired. What started as an epigenetic "sticky note" can sometimes lead to permanent changes in our DNA.

Epigenetic changes are caused by many factors. Sometimes transcriptional processes are involved, so that long non-coding RNAs accumulate. Often there is a connection to metabolism. For example: if there is a lot of a given metabolite, then some enzyme may well use this metabolite to modify a histone tail. Thus, that histone mark would represent or even "encode" an abundance of the metabolite to begin with.

There are many ways how our DNA might be changed; often we refer to pharmacological or even toxicological processes here. UV light can literally melt DNA, causing adjacent thymidine bases to "glue" together. Other covalent bonds can be destroyed by UV light—hence always wear sun protection. Nucleoside-based drugs (we talked about these in a **meandering** in ▶ Sect. 2.4) are used against various types of cancer or viral infections. Acting as "toxic" building blocks, they directly interfere with the synthesis of new DNA. Then there are all sorts of chemicals that either bind to DNA (such as alkylating chemicals) or may even wedge between the nuclear bases. In all these instances, DNA variation is generated at the level of the **individual nuclear base**, such as tpyos in a bigger text.

At the level of **genes** or even chromosomes, DNA is thoroughly mixed regularly during meiosis. And this gene mixing has happened—always; long before the advent of CRISPR gene clippers. Genes can be lost or appear newly in a genome through duplication or even via horizontal gene transfer. Whole chromosome sections can be reversed, duplicated, or even deleted, entirely. Retroviruses or transposons are often involved in these re-arrangements. Silenced and maintained viruses are surely the absolute exception, you might think…? Think again: About 8% of the human genome represents the remnants of old, or even ancient, viral infections. And then we have about 40% of our genome with repeating sequence motifs, which are probably of viral origin. So almost half of our genome, probably is not even ours, but of viral origin?

Humans, animals, and plants are not bacteria or viruses. We have days, weeks, or even years to reproduce—compared to minutes or hours for bacteria or viruses. That means: As long as our viral "legacy" in the genome remains "quiet" and allows itself to be replicated with ease, "we" are happy to carry them along. During this selection, the Southern German Swabian phrase "*Nicht geschimpft, ist Lob genug*" (*not yelled at, is praise enough*) might apply—a neutral mutation or even a silenced virus is sturdily retained, and not commented at, as long as the change is not too disruptive. Some believe that similar mechanisms also come into play in the administration of medium-sized or larger organizations. Only what really disturbs is selected out. Everything that is strange, but not too strange, is consistently retained and passed on.

Compromise The thoughts of Darwin and Lamarck can be translated into modern-day genetic language with the terms "genetics" and "epigenetics". Nature does both—the strict inheritance of DNA from one generation to the next—AND—the fixation of temporary information on the genetic material, initially at the flexible ends of the histones; indirectly, however, this can also lead to DNA changes. Sequence variation occurs randomly and constantly—it is not the exception, but the rule among the genes in any gene pool, between individuals or even between populations. Practical aspects of selection and adaptation are discussed in the following chapter.

▪ Further Reading

What a unifying and forgiving title—Guevara-Salazar et al. **2021**. What are the origins of growing microbial resistance? Both Lamarck and Darwin were right. *Expert Rev Anti Infect Ther*. 19(5): 563–569. ▸ https://doi.org/10.1080/14787210.2021.1839418.

Here I like to quote again the Birmingham Lamarck fan and co-founder of epigenetics Bryan Turner—Turner. **2013**. Lamarck and the nucleosome: evolution and environment across 200 years. *Frontiers in Life Science*. 7:1–2, 2–11, ▸ https://doi.org/10.1080/21553769.2013.835284.

The big evolution anniversary in 2009, the bicentennial of Darwin's birthday, was the occasion for Mr. Koonin to compare thoughts of Darwin and Lamarck on a molecular level. Cheers—Koonin & Wolf. **2009**. Is evolution Darwinian or/and Lamarckian? *Biol Direct*. 4:42. ▸ https://doi.org/10.1186/1745-6150-4-42.

Otherwise, even from the year 2025, you can still find quite emotional comments in publication databases when you search for "Darwin" and "Lamarck" at the same time. I will not list these here. Enjoy reading.

8.3 Practical Evolution: Optimizing Proteins Through Evolution in the Testtube

Regardless of how we interpret Darwin and Lamarck, the creation of variance in DNA is not a big problem—variants are constantly and always arising. So, one only needs to apply simple rules to recreate evolution in a test tube. We "just" need to generate a large number of **protein variants** and then we need to select the "good" variants. Then, we would let this cycle run a few times and the most wonderful protein variants are created, which we would never have come up with ourselves. Oh yes, we also need an **expression system**, in which we can produce the protein.

An expression system, a way to generate variation, and a selection method—these three factors are wonderfully combined in phage display. A phage is a virus that infects bacteria. A rod-shaped filamentous phage can express a foreign gene as a fusion protein with one of its envelope proteins (= **expression system**). Thus, our protein to be optimized is nicely exposed or *displayed*. To manipulate the coding DNA sequence, you can now do all sorts of things: pour mutagenic chemicals in there, perform PCR with loose fragments, or engage with gene synthesis, where very specific positions in the primers are provided with several or all four possible nucleobases. Done properly, DNA mixtures are created with an insane **variance** (maybe 10^6 independent clones or even more). Such a mixture is called a library, probably because of its almost inexhaustible richness. Once transformed back into phages, the entire library often fits into a few drops of buffer solution.

What remains is to sort out the best variants from this mixture (= **selection**). If I am looking for a protein that simply binds to a given target molecule, then I need the target molecule, purified as pure as possible. That target molecule is then fixed to a suitable surface, for example to magnetic beads, and mixed with the library. Some washing steps follow to remove the non-binders from the library. In a final step, the binders are detached, or eluted. This cycle can be done a few times; sometimes sprinkled with additional mutagenesis or shuffling rounds of generating even more genetic diversity.

It is the beauty of this display system that there is a coupling of DNA and the displayed protein. Other **display methods** (for example on the surface of bacteria or even on the ribosome) follow the same principle. All these methods conveniently link the respective protein "on display" and the corresponding coding sequence. The big experimental advantage is, you do not need to analyze the protein directly, which would be very tedious; instead, you only need to sequence the DNA of what remains. Previously, this was a few or a few dozen clones using classic Sanger sequencing; nowadays, entire libraries are sequenced, before and after selection, using next-generation sequencing methods.

The most widely known "binders" are antibodies. And therapeutic **antibodies** are now indispensable in clinical practice—you can recognize antibodies in the "mab" ending of their International Nonproprietary Names (INNs). Monoclonal antibodies represent a huge and ever growing market of very successful and highly effective biopharmaceuticals. Based on a well-known scaffold, antibodies are highly specific binders that can address almost any type of target molecule in the patient with utmost precision. Targets of antibodies usually are molecules that are very difficult to address, or even deemed impossible to address, with conventional pharmacological methods, such as the small molecule drugs aspirin and sildenafil.

As good as they might be as drugs, antibodies are very expensive to produce; our body does not particularly like foreign protein, especially not freely floating in the bloodstream. Therefore, antibodies are quickly digested by the body, chopped into pieces, and excreted—they often do not have a particularly long half-life in the human body. Some other proteins, completely unrelated to antibodies, might become potent binding molecules as well. Even RNA or DNA can be reasonable material for high-affinity binders, these are referred to as aptamers, then. Ideally, all these binders bind exactly what antibodies bind—practically everything. But they all are just common biomolecules and are therefore quickly degraded in the body. We discussed the idea of mirror-image binders already in ► Chapter 2. So much about **biological binders**, for now.

Things get much more complicated when you screen **enzymes** for enhanced activity, or specificity, instead for simply sticking to some sort of a plastic dish. A library is relatively easy to make—the methods mentioned above will still create a good variability. A highlight here is a targeted library, where you chose a number of positions for variability, while leaving the protein backbone untouched. The difficult part only starts then: Each individual variant must be put into an expression system, for example the bacterium *Escherichia coli*. Each protein needs to be expressed and purified. Finally, it must be examined in a suitable assay system for the desired properties. Even if all these steps are adopted to 96-well plates, it remains a lot of donkey work, even to just analyze some hundred variants. Regardless of entire libraries been severely "under-sampled", yet astonishing improvements in enzyme properties are achieved every now and then. The field is moving fast—it gets exciting when methods start to merge; people have already packed enzymes into small droplets for FACS, or fixed them on high-throughput sequencing chips, for highly parallel imaging.

Collection Predicting the effects of a single point mutation on the behavior of an enzyme is difficult. Predicting the course of a large-scale selection experiment of protein evolution (still) is absolutely impossible. Display systems can link a protein to its coding nucleic acid (be it DNA or RNA). In such a system, variance can be generated in all sorts of clever molecular biology. All that remains is to select suitable variants; but this selection often is the limiting factor, as entire libraries of 10^6 or even 10^9 variants can only be fully examined with very clever and/or highly technical screening methods. Evolution-in-a-test-tube has made astonishing progress both in biomedicine and in industrial enzyme technology—please think of our dishwashers and washing machines; think of chemo-enzymatic catalysis, of excellent detergents, and of intricate binding molecules that dock highly specifically to biomedical target structures.

▪ Navigation

We have already mentioned phage-display in a meandering in ► Sect. 2.6. The discussion about DNA in ► Sects. 3.2 and 3.4 might also be interesting, for a second read.

How to Read, and Write, Scientific Publications

Contents

© The Author(s), under exclusive license to Springer-Verlag GmbH, DE, part of Springer Nature 2026
J. W. Mueller, *Ultimately Understanding Biochemistry*, https://doi.org/10.1007/978-3-662-71889-6_9

9.1 Different Scientific Languages Make it Difficult to Read Scientific Papers

In this book, I could have addressed many things. Scientific methods, techniques, countless sub-disciplines and so many colorful facts about a myriad of enzymes. You can learn all this, however, once you understand the language of science. This is why I dedicate this last chapter to reading papers, written in said scientific language; and maybe even to consider writing your own paper. Learning any new specialty—be it a scientific discipline, or a bespoke kind of sport like yoga, or sailing on a yacht—is pretty much like learning a new language. Generally speaking, we have to master not just one, but at least three different languages, in experimental sciences; and all of this without considering any specialization, yet... The three languages are:

- A somewhat cryptic and technical language (**lab-slang**),
- The **language of papers** and publications (objective and neutral) and
- A conference and **proposal language** (flowery and euphoric), the language of press releases, of outreach events, and posts on social media platforms.

Lab-slang is full of incomprehensible abbreviations and technical terms, something like "we have just observed ABC with method XYZ, we have no idea what ABC is supposed to mean, but we write it down, regardless, just for us, in our lab, in silence..."

The **language of scientific publications** is neutral and objective, in the sense of "we know very well that this what we are reporting, is fabulously groundbreaking, but please just look at our pretty data, and make up your own opinion". Traditionally written in perfectly modest understatement; one might even read a somewhat *British* sentiment into this. A fantastic example of this language is the last sentence of the iconic 1953 publication by Watson and Crick, on the structure of DNA:

> "*It has not escaped our notice that the specific pairing we have postulated immediately suggests a possible copying mechanism for the genetic material.*" Watson JD, Crick FH. A structure for deoxyribose nucleic acid. Nature 1953; 171:737–738.

This is it! This is the WHOLE and entire discussion of one of the most sensational papers, ever!!! Less is more. Roughly translated for the general audience [and expanded by JWM]: "We have certainly noticed that the model IMMEDIATELY explains the mechanism of DNA replication. It, much more, lays the foundations of modern molecular biology [but the reader has to think for themself. If he or she can't do this, then we ask them to please wait until our next suite of papers comes out]". This second half of my extended transcript subtly resonates with the first one in British English, and, at some later occasion, both Watson and Crick have indeed published on the matter, extensively.

Almost independent of all of the above is the **language of press releases** or conference contributions. This language is also frequently found in research **proposals**. All of it sounds like: "What we have just discovered is at least (!!!) as great as the invention of sliced bread. Further investigations are urgently required."

Unfortunately, these different languages are muddled together, occasionally. So, you will often encounter lab jargon—technical vocabulary, rich in abbreviations and technical terms—where it actually doesn't belong, be it within publications or funding applications. Trying to make sense of documents with such mishmash often is difficult for scientists and students alike.

> *"There is no form of prose more difficult to understand and more tedious to read than the average scientific paper."*
>
> Francis Crick—The Astonishing Hypothesis: The Scientific Search for the Soul (1995)

When reading written assignments from students, the author of this book sometimes gets the impression that students are keen to use the very weapon that usually prevents them from understanding scientific publications—maybe to try to pay back? Some pretend to just wanting to reduce their wordcount… Whatever the motivation, the one or the other thesis contains many unnecessary **abbreviations**. They might only be used once or twice, but the damage is done. Such abbreviations are not necessarily helpful in reading the piece quickly. And if no one can read or understand your work, any arguing about saving space is obsolete. Regarding the use of abbreviations, the honorable *Journal of Biological Chemistry* is my role model: There you can use abbreviations from a general list that includes DNA, H_2O, and a good couple of other common abbreviations. But if you introduce a new abbreviation, you have to use it ten times or more; otherwise, you will need to spell it out nine times or less.

ℹ Personally, I would increase the rule around abbreviations to a dozen! But what is your opinion? Do you know publishers or magazines with similarly strict rules regarding unusual abbreviations like the *Journal of Biological Chemistry*?

While it sounds like a nice idea to distinguish between general and specific abbreviations, I had to learn that this is a perfectly personal thing, depending on one's background. Our students tend to have great difficulty distinguishing a "universal" abbreviation from one that is only valid in their "own" field.

How to give general advice here? Perhaps this rule might help: Everything that was taught in your course of study should at least be generally known to your cohort of students; maybe also count in your supervisors and potential markers. Let's call those "general" abbreviations then. Everything that has to do with your thesis project itself, may only concern one or two students—so be especially careful with these "subject-specific" abbreviations. Btw, do you actually know **TLA**s? I might also want to comment with **IDUYA**, at appropriate places. Confused? Please read the next paragraph for some framing…

The purpose of using the **TLA** and **IDUYA** abbreviations without introducing them, was to show how dangerously unintelligible abbreviations can be. Without context they mean nothing to any of us. Anyone with a serious interest in the origin of the above abbreviations, is reminded of English *t̲hree-l̲etter-a̲bbreviations* (**TLA**s) and the saying: ***I̲ d̲****on't* *u̲nderstand y̲our a̲bbreviation* (**IDUYA**). My allergy about jargon and special abbreviations has paired up recently with a great skepticism towards "small words" or pronouns—just to be clear, pronouns in email signatures serve a different purpose, absolutely fine. In the middle of scientific texts, however, it is not always clear who "it" is, in this case, or what "this" represents, in that case…

Understanding scientific publications, can be problematic not only on the level of the actual scientific text. The language around publishing can be cryptic as well—but, deeply into it, we ourselves often no longer realize the problem. For first-year students, it might NOT be obvious that a "figure" is not a number, but some sort of graphical representation of scientific data. More related stuff is shown in (◘ Fig. 9.1).

I like to compare **reading scientific papers** with **reading newspapers**, or at least scrolling through a news app on your phone. Very, very rarely would you read a newspaper from beginning to end. Rather, one flips through the pages, or scrolls up and down on the app, and scans the headlines until you find a paragraph or article

9

◘ **Fig. 9.1** **Common misunderstandings around scientific publications**. **Top left** is "Fig. 1" to be seen, the first fig of the year from our garden. In seminar settings around scientific publications, students sometimes look at me bamboozled, when I recommend them to pay particular attention to the figs... To what? **Top right**, you might see a fig legend. **Bottom left**, that clearly must be a summary for a "lay person", so a layman or a lying person? **Bottom right**, could that possibly be the abstract of your paper? Not shown is one of the most productive scientists ever. Maybe you have heard of Professor Et Al, as well. That academic is mentioned a lot, everywhere. Do you know where? (Image credit: Fig, JWM 2020; Fig-Legend, JWM 2021; Lay summary, JWM, 2022; Abstract, John Gage, 2021, with kind permission)

that seems relevant or interesting. Only then to read that specific part somewhat more focused, or—more likely—to bookmark or to like it for later.

What exactly is relevant in a scientific paper, for any specific student or researcher, depends heavily on their respective scientific background, research interests and their current career stage. As a student, it was important for me to somehow understand what was **depicted** in the **figures** and how the paper endorsed these findings. More specifically, this meant comparing the authors' claim with the provided data, tables, and figures from the same publication to support their claim. Our students regularly struggle in analyzing and interpreting these figures.

During most of my doctoral thesis, a lot was about which **methods** others used in their paper. Would I be able to understand these methods; would I even be able to use them in my research project? In the last year of my doctoral thesis, the **address line**—the affiliation—became really interesting to me. Where on planet Earth was this study conducted? Could this possibly be a good lab in an interesting part of the world, for postgraduate research, for doing a postdoc?

Always, I regarded the list of **authors** quite interesting. One can immediately look up what these ladies and gents have published, in addition to the paper in front of me? Have I met any of these people, maybe in a local seminar or at a conference, already? All scientific researchers care about how often their work has been **cited**, as this greatly influences their scientific career—at the latest since the advent of the Hirsch factor (please look it up in the glossary). I have heard of people who first check in any given publication whether they themselves were cited; would you believe it...

After all these years, I nowadays focus on understanding the figures and comparing them with the statements and claims of the paper. I always pay attention to which lab has published the paper and what methods were used. Along the way, I also check whether the authors have not forgotten to cite relevant publications, including but not limited to our own research, just saying.

Clarity **No one** reads a scientific paper from A to Z, NO ONE! For different purposes—to give a presentation, to use a method from the paper itself or even to build on data presented in that paper—different parts of a paper are more important than others. Therefore, your approach to a scientific paper should be adjusted accordingly.

9.2 ...and How Should One Write an Easy-to-Understand, Widely-Read Paper, That Is Highly Cited...

There are many guides to scientific writing. In addition to all of them, I can only recommend that you try to put yourself in the reader's shoes, every now and then. Maybe try to look back after you have written your piece; maybe even after you haven't touched it for two or three days—how does your "cured" manuscript look like then, from the perspective of the reader? It is then that you probably would be able to look at your own manuscript with fresh eyes. More pragmatically, this section is to convert a good number of ideas from the last sections, into recommendations, to entertain the examiner, the reviewer, and most importantly, the interested reader, out there...

Please be **very, very careful with** your most bespoke special **abbreviations**, for the sake of the argument, let's abbreviate them today as **YMBSA**, intentionally or carelessly different to what we introduced just above. If you do not use these abbreviations often (say, ten times or more), it is better to write them out each time. If you got lost, we are still talking of **YMBSA**, **y**our **m**ost **b**espoke **s**pecial **a**bbreviations.

Using pronouns in scientific texts can be a good strategy to avoid too frequent word repetition. Please, however, be always clear about what or whom you mean. How about a kind of a knitting pattern: Name the thing (e.g. Dörk3) and immediately group a second term with it (the kinase Dörk3), then use a conjunction, then again Dörk3, then kinase, then conjunction, etc... This ensures both variety and clarity. Interested in Dörk3? You can read more about this special kinase in the **Epilogue** section.

Most likely, **the reader will NOT read your paper from A to Z**. That's a fact. So, you better design each section to be largely readable and understandable on its own. Cross-referencing to other sections is allowed, as long as these links are informative, but do not hinder or interrupt further reading. An absolutely waste of words is a generic reference such as "As mentioned above, …". It only steers bad feelings, as you press me to admit that I might have forgotten was there was up there, already…

Why not reintroduce your most bespoke special **abbreviations** (YMBSA) at the beginning of each section? Nothing is celestially evident—absolutely nothing is clear to all of us, at all times. Things as seemingly clear as "ER" can stand for endoplasmic reticulum, or the estrogen receptor in biomedical research; it can be the emergency room in case reports in a clinical setting; or the chemical element erbium in more bio-inorganic analytical applications. And we have not yet started with Royal Ciphers. **YOU** as the author **need to define your abbreviations**—if you leave this to the audience, you might create great misconceptions.

Ideally, write the paper that you would like to read by yourself, be it as a student, as a scientist, as an examiner or even as an evaluator during peer-review. For sure, your writing style will improve if you read your text while putting yourself in a few of other roles, such as the editor, the referee, or the interested reader. Oh, and then there is the winning pattern around Shakespeare's writing style. He is said to have divided his works into three parts—about a third of each work was for the rural population—flat jokes, crude innuendos, that peasants could understand easily. A third was for the civic class—intrigues, relationships, etc... And a third was for some highly educated aristocrats and peers—attractive prose, fine rhymes, and historical reflections. Why can't we structure our papers in the same way—crystal clear essentials for everyone—supported by a really good graphical abstract, understandable facts for those who are interested in it a bit more, and a few somewhat encrypted facts for fans and science geeks. Would that be an idea?

Commentary **Very few scientific papers are read from the beginning to the bitter end**, in their entirety. How one should read a paper depends crucially on what one intends to do with it. Depending on where you are in your scientific career, different sections of a paper might change in their importance to you. Carelessly used abbreviations or unclear sentence structure, will massively hinder others to read and understand your paper. This mindset should influence your approach towards writing a paper. Please be very careful with abbreviations and "small words" and also design individual sections easy to understand on their own. A **meandering** provides an impressive example of good and bad papers.

Meandering: Of Well-Written and Badly-Written Scientific Publications

I apologize for returning to the race around DNA. It is a prime example of very well and very poorly written papers. Some people may think that Watson and Crick rushed ahead with their paper. Franklin was pushed to the side. And Wilkins' sad role was to simply confirm the DNA model **later**. **This is not true**. At least not for what happened in 1953.

On April 25, 1953, the reputable journal Nature published **three** papers on DNA. We all may have heard of the publication by Watson and Crick. Back-to-back-to-back, Rosaline Franklin and her doctoral student Gosling, as well as Wilkins with his colleagues Stokes and Wilson, also published their experimental data. Ta-dah, the dispute between Franklin, Wilkins, and others, was resolved for the benefit of all. **Everyone published at the same time**, everyone had their Nature paper, everyone was happy. Franklin was NOT left behind.

History tells a different picture. I have never heard of the Watson-Crick-Franklin-Gosling-Wilkins-Stokes-Wilson base pairing or a similarly named DNA structure. It was only Watson and Crick receiving widespread attention for their clear and catchy model. On the occasion of the 50th DNA anniversary, Martin Tobin compared the three papers for their content and their writing style.

The paper by Watson and Crick is short, it is only 842 words long. The text is written clearly and understandably. No sentence has to be read several times in a cumbersome way to understand it. **The only abbreviation that the authors introduce is "DNA"**. The other two papers, though, are about twice as long. They are overloaded with cryptic-technical jargon. Even though these two papers contained the actual experimental data, they seemingly were forgotten quickly; maybe in part because of their poor presentation. What has not been forgotten, however, is the super cool final sentence of the Watson-Crick paper, which we have already discussed at the beginning of this chapter.

The model of Watson and Crick, which made its way into the textbooks, was based exclusively on experimental data from the other two publications, and the two authors rightfully have been criticized about this fact. The authors of those other articles, however, "missed" the entry into the history books probably also because they could not express themselves briefly and concisely.

This meandering is about clear writing and accessible communication. The characters involved, however, have been the subject of a lot of discussion you can read about it at some other place. Please note however that Watson only passed away in late 2025. I read quite a number of obituaries about him; I only want to recommend one outstanding piece that Sharon Begley had written about James Watson before she herself died in 2021.

Those three papers...
Watson JD, Crick FH. A structure for deoxyribose nucleic acid. Nature 1953; 171:737–738.

Franklin RE, Gosling RG. Molecular configuration in sodium thymonucleate. Nature 1953; 171:740–741.

Wilkins MHF, Stokes AR, Wilson HR. Molecular structure of deoxypentose nucleic acids. Nature 1953; 171:738–740.

...scrutinized by:
Tobin MJ. April 25, 1953: three papers, three lessons. Am J Respir Crit Care Med. **2003**; 167(8):1047–1049. Editorial. ▶ https://doi.org/10.1164/rccm.2302011.

...and, a very special remembrance piece—Begley S. 2025. James Watson, dead at 97, was a scientific legend and a pariah among his peers. In the Lab. Remembrance. StatNews. ▶ https://www.statnews.com/2025/11/07/james-watsonremembrance-from-dna-pioneer-to-pariah/ [accessed 7th March 2026]

9.3 Well, There Is One Last Thing: English as the Language of Science

Until here, I have pretended that learning one or several "scientific" languages would be enough to study your preferred discipline. For most of us on this Planet, learning English as a foreign language definitely needs to be added to this list. This sub-chapter it dedicated to all my non-native English speaking colleagues, who like to study biochemistry.

As of now, the **language of science** is still English, it functions as a Lingua franca, a vehicular language. Scientific English is not that difficult if you have understood the science it tries to describe. This makes reading and writing scientific texts relatively easy, even for non-native speakers. Texts in "scientific language" are often relatively easy to understand—so much easier to understand than prosaic texts by, perhaps, William Shakespeare, Jane Austen or Alan Hollinghurst. As long as you have acquired the scientific vocabulary, there is really nothing to be scared of.

One of my university lecturers told us that we should use English textbooks for revision from as early as possible. They are usually one edition ahead, or two. Often, they are even cheaper than the translations. And by the way, you also improve your written English. A *Win-Win-Win* situation. With academic experience in Germany, and the UK, I can only endorse this statement.

Having sorted the "written" part, what remains is the **verbal part** of scientific English. It is crucial that you understand your conversation counterpart adequately. Unfortunately, understanding can be very difficult, at the beginning; especially if conversations take place in noisy surroundings, such as the airport, a restaurant, or a side room crammed with posters and people, in a hotel. But I promise, things will really get better after a while; keep trying. A cup of tea or a pint of beer might help in finding common grounds for understanding.

And then comes the part where you have to answer. I look back on quite some years of English experience, having lived in the UK for more than 15 years. I have organized more than a handful of conferences myself; and have attended many more scientific meetings. It might come with disappointment, but if you have a certain accent, it will probably never disappear, unfortunately. I personally have the impression that my own German accent will never disappear as long as I speak German on a regular basis, for example with my partner at home, or with the dear relatives back home, in Germany.

So, it is probably best to learn to love your accent. Don't overestimate your learned language skills, as your accent still will provoke that special question, you know, which one: "Where are you from…" Followed by this one: "No, where are you actually from…" I'm able to take this all with a smile, meanwhile, explaining that our family is from around Berlin and from around Cologne, and then just listen to the experiences of the other side might have about Bavarian Oktoberfest, from Heidelberg, or Osnabrück and Colditz, from back in the days… Let's deeply hope that no anecdotes come up on topics like "*What-did-we-have-a-great-war*" …

One of my linguistic **highlights** was a certain response to one questions of my faculty-wide teaching evaluation: "*What aspects of Dr Jon Mueller's approach to teaching best helped your learning?*" Answer: "*His accent. It helped me to revise because I would try to mimic his soft German accent, in my imagination.*"

One of my absolute **lowest points** in stammering in English: At a conference in Tel Aviv, Israel, I met the famous crystallographer Ada Yonath, only to greet her mumbling something like this: "*I'm so… so… happy to see you alive*" [probably also with a harsh German accent]. Her response: "*I'm also very happy to be alive*", and she moved on...

Communication The primary scientific language has changed over the centuries. After it was Latin in the Middle Ages, German and French were quite common at the end of the 19th/beginning of the twentieth century. "We" Germans then did a lot of foolish things to abolish German as a standard common scientific language. Leaving us with American English as *Lingua Franka* for now. How might the future look like? English will probably remain the number ONE scientific language for a while before it is replaced… maybe by Mandarin Chinese, maybe by multiple languages and much-improved machine translations… You never know.

Navigation

A thought-provoking article about whether the input of a (non-specialist) native speaker helps in scientific writing, or

not—Pawlicka et al. 2022. Opening up a discussion on the native speaker myth in science. *Account Res.* 29(7):477–481. ▶ https://doi.org/10.1080/08989621.2021.1960514

Irony may convey key messages better than the truth. Hence, here is a guide on how to make sure your writing guaranteed is incomprehensible and boring—Sand-Jensen. **2007**. How to write consistently boring scientific literature. *Oikos. 116* (5), 723–727. ▶ https://doi.org/10.1111/j.2007.0030-1299.15674.x.

A somewhat more classic guide to good writing—Gemayel. **2016**. How to Write a Scientific Paper. *FEBS J.* ▶ https://doi.org/10.1111/febs.13918.

Please also check out this recent guide by a Birmingham colleague of mine—Roberts. **2024**. Advice to a Young Mathematical Biologist. *Bull Math Biol.* 86(5):52. ▶ https://doi.org/10.1007/s11538-024-01269-1.

The author of this book briefly tried to write guide literature. Here are a few tips for successful rebuttal writing—Mueller. **2021**. A rebuttal is that letter that you send to a scientific journal in response to the loving and helpful comments of the expert reviewers. *ASBMB Today*. ► https://www.asbmb.org/asbmb-today/careers/102821/how-to-reply-to-reviewers-comments. [accessed 28. April 2024].

...and an interesting opinion about Scientific English and the myth around native speakers—Pawlicka et al. **2022**. Opening up a discussion on the native speaker myth in science. *Accountability in Research*. 29(7):477–481. Commentary. ► https://doi.org/10.1080/08989621.2021.1960514.

Supplementary Information

© The Editor(s) (if applicable) and The Author(s), under exclusive license to Springer-Verlag GmbH, DE, part of Springer Nature 2026

J. W. Mueller, *Ultimately Understanding Biochemistry*, https://doi.org/10.1007/978-3-662-71889-6

Epilogue: Favorite Molecules of a Lifetime

Nature has always fascinated me, with its rich shapes and colors. My father was an impressive role model for me, a confirmed scholar brooding over his books. At the sweet age of nine or so, I insisted on dressing up as a professor for the school costume party. Believe it or not, but back then I had to use a pillow to indicate a beer belly. At about eleven years of age, I dismantled a battery, to electrolyze a sodium chloride salt solution with the carbon electrodes that I just had stripped free, delightfully enjoying the refreshing scent of the ascending chlorine fumes.

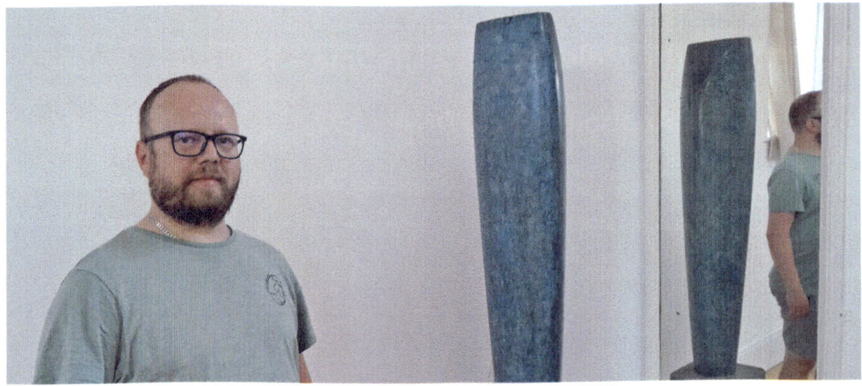

Fig. 1 The author at the Barbara Hepworth Museum in St. Ives, UK; next to a gorgeous stone sculpture. A mirror allows some backside view of the scenery. (Photo: Oliver Schmitz 2021; Reproduction with kind permission)

My chemistry courses at school started when I was 13 years old. For the topic "oxidation", each pupil was allowed to burned their own chip of magnesium—it happened that my scorching hot magnesium piece dropped, burning a nice brand mark into the table. At university, calixarenes were very interesting to me in my first year, with their use as host molecules in supramolecular chemistry. But then in my second-year biochemistry lectures, and I fell in love with TIM. TIM, also known as triose phosphate isomerase, is a perfect enzyme. TIM is so elegantly fast that only the diffusion of substrates and products limit this enzyme, that's very fast. Moreover, TIM is also good-looking—a central barrel, formed of eight β-strands, decorated by eight α-helices all around the barrel.

Next on my list was a thiamine pyrophosphate-dependent enzyme—I was fascinated by yeast pyruvate decarboxylase. This enzyme not only is central to making good pizza dough, it also is crucial in the formation of alcohol. Very likely, pyruvate decarboxylase is the enzyme that I have learned about and worked with, with the greatest personal commitment. Then came the time of my final year thesis project, which I completed in cooperation with an international pharmaceutical company. The molecule of choice was called a target now—it was the protein kinase DYRK3, "Dörk3", that was functionally linked with hematopoietic signal transmission. DYRK was hard work. During this time, I probably learned less about enzymes, but all the more about frustration tolerance.

Asking recombinant bacteria to make DYRK, only resulted in greasy lumps of slimy bacterial pellets. Quite unlike some happy small folding helper proteins, which we could produce easily in large quantities. During my early postdoc time, I researched two representatives of these so-called peptidyl-prolyl *cis/trans* isomerases. There was the star protein Pin1, a folding helper for phosphorylated peptides. Pin1 was insatiable—it played a role in almost every process you looked at: a master regulator? Maybe a head chef? And then there was the small commis chef protein Parvulin 14, and a slightly longer version 17, named for the Latin word "*parvulus*" which means tiny. Our research deepened the understanding of the cellular processes these enzymes are involved in. We even highlighted core features of the actual catalytic mechanism. What these folding helpers actually are good for, however, remains a mystery, until this very day.

Sometimes it seems like a good idea, not to have all the eggs in one basket. Whatever we learned about the little folding helpers, colleagues around me looked at enzymes that attach a sulfate group to tyrosine side chains of proteins. Their attempts at producing these proteins were at least as frustrating as my work with DYRK. However, these sulfotransferase enzymes had to be supplied withan activated cofactor for sulfation—let's call it PAPS for now. And within human physiology, there are only two enzymes, the cousins PAPSS1 and PAPSS2, that can make PAPS. It was my love for these PAPS syntheses that eventually brought me to England. An important part of our work here is to study patients whose genes for those PAPS producers are non-functional. I have been thinking about form, function, and significance of the PAPS cofactor itself for a while. I look at "my" sulfation pathways from within endocrine biochemistry not only as researcher, teacher, and textbook author, but also as conference organizer and editor of special volumes from a somewhat broader perspective.

Did I mention that I think that biological sulfate esters are quite exciting molecules—especially if they involve steroid hormones? This research interest persists until this very day. On top, I have also started to search for challenges—is there a really strange protein around, that one could use all ones biochemical experience at? The answer is: Yes, there is plenty of stuff around us that still needs to be studied.

Glossary: From Anabolism to Z-DNA

Would you like to have more technical terms explained my way? Please let me know.

3D-Chemistry → Stereochemistry

Anabolism: All substance-generating processes taken together—biosynthetic reactions that produce the building blocks of the cell. A typical anabolic reaction is a reductive biosynthesis in our body, where ATP and NADPH are used, to build biomass.

Ancient DNA, old DNA, very old DNA: DNA is very stable. Therefore, DNA can be isolated and sequenced from almost any old material, with only moderate effort. Probably a matter of debate is what "old" exactly is. Most likely, one can analyze the DNA of people who died in the first half of the twentieth century. Under lucky coincidences, ancient DNA even might be sequenced from a long-extinct species, say 500,000 years ago. Instead, it definitely is not possible to sequence 65-million-year-old dinosaur DNA, be it included in amber or not. DNA is not that stable.

Angstrom (Ångström): The Ångström (1×10^{-10} meter) transforms any measure in molecules back into practical numbers. Thanks to the unit Ångström, the diameter of a hydrogen atom is no longer 1.1×10^{-10} m, but 1.1 Ångström, which is so much nicer to say, isn't it… The Ångström is named after a Swedish physicist of the same name who dealt with atoms a good while ago.

Antibody: The binders that existed before aptamers. Antibodies are an important part of our immune defense. Immunologists group them in various classes. The structure of a standard antibody vaguely resembles the letter Y, a wye, an *ypsilon*, where the top two ends contain highly variable binding sites. The diversity at these binding sites is so massive that there probably is an appropriate antibody for ANY given binding motif. The antibodies of most mammals consist of two light and two heavy proteins, also referred to as "chains" here. The antibodies of camels and llamas are a big exception—they consist of heavy chains only.

Aptamer: A fairly new word that refers to any binder to a specific target molecule. An aptamer can consist of any biopolymer. RNA aptamers are most common; you'll find DNA aptamers less often; some say that even peptide aptamers exist. "Apta" comes from the Latin "*aptus*" which means fitting. Probably the closest we get to this word stem is the word Ad-**apt**-er. The syllable "-mer" at the end, means "part" of something (→ polymer).

Aromatic: Aromatic compounds are a group of organic compounds with a ring of carbon atoms, all linked exclusively by double bonds. The ring can have five or six corners. Non-carbon atoms, such as oxygen or nitrogen, are happily incorporated here. Aromatic compounds were named for the aromatic smell of the first discovered representatives. Probably a misleading name—some isoprenoids and reaction products of sugars with amino acids smell much tastier, even though they are not aromatic.

Bond, **Anhydride bond**: Some acids are oxo-acids—they contain at least one oxygen atom. Let's think of phosphoric acid here. When you force two such acid molecules very close together, they will fuse, while kicking out a molecule of water. The two acid residues are now connected via an oxygen bridge—let's think of pyrophosphate here.

Bond, **Double-ester bond**: Phosphate, it is again. Phosphate actually has three bonding possibilities. Two of them are usually given to sugars (in RNA and DNA). This creates a situation where a negative charge is literally "hovering" over the double ester bond. This additional negative charge stabilizes the double-ester substantially, as it prevents nucleophiles, such as water, from approaching and breaking one of the ester bonds.

Bond, **Ester bond**: An acid and an alcohol can form an ester bond, by splitting off water. In biochemistry, this bond is popular and widespread—it is easily formed, reasonably stable, but also quite easy to cleave again. The acid usually is a carboxylic acid, but sometimes it can be something else, such as phosphate or sulfate (or formally their protonated acids). The alcohol basically can be any substances with a hydroxyl groups. Thio-alcohols are a notable exception, where it is a thiol group (-SH) that bonds to an acid. In biochemistry, the most prominent representative of thio-alcohols is coenzyme A.

Bond, **Glycosidic bond**: When you look at glucose in its six-membered-ring form, all OH groups look the same. One of the six, however, is "more equal" than the others—and that is the OH group at the C-atom #1. This hydroxyl is not a real OH group. Instead, it is an aldehyde group in disguise, only visible as aldehyde, when glucose converts into its elongated form. When this OH group bonds with an alcohol, an O-glycosidic bond is formed. If it engages with an amine (a nucleobase maybe) an N-glycosidic bond is formed.

Bond, **Hydrogen bond**: We biochemists love talking about hydrogen bridges. An absolute highlight of biochemistry. The H-bridge is some sort of mixture of a polar bond and a little bit of covalent bond. There are H-bridge donors and acceptors. In biochemistry, these donors and acceptors nearly always consist of hydroxyl and/or amino groups.

Buffer/Buffer solution/Buffer system: Biomolecules, be they small or large, that can absorb or release protons, function as buffers at about "physiological" pH values. Buffers mitigate changes in pH value. Most compartments of living things are carefully pH-controlled.

Catabolism: All reactions from which the body can derive energy, taken together. Catabolism also includes molecular waste disposal and proper recycling, at times even down to the individual atom.

Chemical space: In this book, this concept is mainly used to illustrate how much we actually NOT cover here. Otherwise, this concept is probably mostly used when screening huge substance libraries. This way one can estimate how bad the under-sampling in your experiment is. Most of the time, under-sampling of chemical space is very, very bad—only a tiny part of the huge complexity of many libraries actually takes part in a lot of experiments. Related concepts are **sequence space** of DNA or **structure space** of proteins.

CHNOPS elements: Isn't this a nice abbreviation for the essential macro-elements carbon, hydrogen, nitrogen, oxygen, phosphorus, and sulfur. Life on Earth needs all these elements in abundance. Life also needs quite a few more elements, but those are only needed in tiny quantities—they are the micro-elements then.

Coenzyme → Cofactor

Cofactor: Cofactors are important for a lot of enzymes. Many cofactors are made of a vitamin component, and some sort of nucleotide; amino acids are rarely used as well. A cofactor can freely diffuse around or be bound to the enzyme. Cofactors often allow a special sort of chemistry—this may include a cofactor to tightly hold small and "slippery" substrates, acting as some sort of pincers; cofactors might also activate somewhat larger "lazy" substrates, such as sulfate or a carboxylate, by binding them. Examples of cofactors are ATP and coenzyme A. This textbook does not distinguish between cofactor, cosubstrate, and coenzyme, but other textbooks probably do.

Color indicator → Indicator

Computer simulation: Maybe trivial, this is some part of life that the computer imitates. In biochemical everyday research however, computer simulations have become a central piece of scientific research. Simulating biomolecules with thousands or even millions of atoms, still brings even the very best computers to their limits. So, until this day, simulations of larger systems are usually confined to some fractions of a second.

Concentration difference: It is quite astonishing, how many strange things can serve as energy sources for life. Concentration differences between neighboring areas of almost any kind, can drive biochemical processes.

Cosubstrate → Cofactor

Dalton, abbreviated as "Da": The Dalton is a unit of the actual mass of an individual atom or a molecule, used commonly in biochemistry. Other disciplines may use "u" instead. The molar mass in gram per mole—g/mol—is a related concept, though not identical. The molar mass refers to the mass of a whole mole of individual molecules, which is about 6.022×10^{23} pieces. Yes, that's quite a lot of particles.

Density anomaly: Most materials expand when heated and contract when cooled down. A few compounds behave strangely different, however. Out of these, water is the most important representative for life. Water reaches its highest density at 4 °C; but expands again below this temperature. This is why ice floats on water.

Dielectric constant: Difficult concept. It doesn't really help that it has been renamed to (relative) permittivity, lately. The dielectric constant refers to the chemical polarity of a solvent. At room temperature, the dielectric constant of water is 80.1; the one of hexane is 1.9. Because of its extremely high dielectric constant, water can dissolve polar substances so well.

Dimer → Polymer

Display methods → Display

Display, **Phage display**: The showcasing of any biomolecule, usually proteins or RNA molecules, on the surface of a bacteriophage. At the same time, the coding sequence is cleverly packaged within the phage particle. There are different packaging systems for displaying biomolecule, in addition to bacteriophages: One could display something on an entire *E. coli* bacterium, a yeast cell, a ribosome, or something else, that is robust enough and well characterized. For molecular evolution experiments, one may want to create variation in the genetic information, encoding the displayed biopolymer. That mixture of variants is understood as → Library.

Dissociation constant: A measure for how strongly, or weakly, two biomolecules are attracted to each other. Actually, the dissociation constant just is a normal equilibrium constant for a binding reaction A + B ⇄ AB, a biochemical complex formation. But since life is difficult enough, one always looks at the reverse reaction AB ⇄ A + B—the dissociation of the complex—which has the much nicer unit "mol per liter", instead of $[\text{mol per liter}]^{-1}$.

DNA variation: Genetic diversity is the building material of evolution. Many different genetic, chemical, environmental and toxicological processes are involved, to create diversity. Changes in individual base pairs can arise spontaneously; they are one type of mutation. If the variation on the nucleotide level is widespread in a certain population, we call them polymorphisms, nicely abbreviated as SNP (*single nucleotide polymorphisms*). → Mutation → Library

Double-ester bond → Bond

Egg white: An abundant and easy-to-access source of protein is egg white. In fact, so common that "egg white" or "albumin", which is a Latin version of "egg white" was used as a substitute for "protein". At least German biochemists use "eiweiß" quite a bit, instead of protein, a questionable habit that partially persists until today. → Protein

Enantiomer: An enantiomer pair are two molecules that are "opposite" (at least that's what is in the word "enantio"). In biochemistry, an enantiomeric pair are the two versions of a chiral compound, they behave like your left and your right hand. Biochemists usually refer to the two as D- and L-form. Chemists prefer the R- and S-nomenclature.

Essential macro elements → CHNOPS

Ester bond → Bond

Expression system: It is so important to be able to express yourself well. But that's not what is meant by an expression system. Rather, it's about producing proteins (or other biomolecules). This can become arbitrarily difficult, as proteins can behave highly "individually". A simple expression system is our lab pet *Escherichia coli*—insert plasmid with the coding sequence, let *E. coli* grow, tickle a bit (induce), let grow further, harvest, break cells, purify protein, done. Protein production works somewhat similarly in various yeasts, in insect or mammalian cells, in transgenic animals or cell-free extracts. Each of these systems has advantages, but also pitfalls.

Flow equilibrium: An "open" system that has a constant throughput of nutrients and/or energy. Nevertheless, such a system reaches a stable state, a *steady state*. Most living things can be described as such a flow equilibrium.

Frozen accident → Prebiotic chemistry

Glossary: Somewhat self-referential to have that term in here as well. It reminds me of the fact that the word "google" is among the most googled words at Google. Presumably because people don't realize that the "naked" navigation bar of most browsers is already an interface with some search engine.

Glycosidic bond → Bond

Heat capacity: Each material is somewhat unique when it comes to whether it wants to be heated easily, or not. Metals usually feel cold or hot only briefly; they are easy to heat up, or to cool down. Water instead is a huge heat sink.

High-energy compound → High-energy molecule

High-energy molecule: A somewhat bulky concept. Biochemically, it probably refers to anything that is somehow capable of turning ADP into an ATP molecule. However, the author of this book finds it important that it is NOT about a single "highly charged" bond; the entire molecule is always involved in the reaction.

Hirsch factor: Also known as h index. One of many ways of quantifying the research outputs of scientists. If you list all of your papers from the most-cited to the least-cited, your h index is the number of that paper that has the same or more citations than that number. In late 2025, my h index was 24. The index is debated, as it favors certain behaviors, such as h-index-bashing, or just getting old.

Hoogsteen base pairing: Our DNA contains the four nucleotides A, C, G, and T. A binds to T, and C, to G, in what is referred to as Watson-Crick base pairs. This leaves a good part of these nucleo-bases void of interaction. And it is these alternative interactions, investigated by some Mr. Hoogsteen, that now are called Hoogsteen binding.

Hydrogen bond, also Hydrogen bridge → Bond

Hydrophobic effect: A biochemistry classic. Just because water does not want to be ordered—entropically speaking—it presses molecules together that refuse to dissolve in water. Somehow an indirect force, but an important one.

Icosahedron: Viral protein shells often have the shape of an icosahedron; some giant protein complexes do too. Just like tetrahedra and octahedra, an icosahedron is made up of equilateral triangles, but not four ("tetra") or eight ("octa"), but twenty ("icosa"). Probably better-known than the icosahedron itself is a thing where the corners are cut off—the truncated icosahedron, the well-known football.

Indicator: Indicators, reporters, sensors, all these are compounds or biotechnological constructs that are supposed to indicate the presence or absence of something. Color indicators are well suited to show changes in acid-base behavior. My favorite pH indicator is definitely red cabbage juice.

Intracellular messenger → Second Messenger

Isoform: Two isoforms of a protein or another biomolecule are similar to each other, but not identical. There's nothing more to the word. Instead, there is ambiguity—it could refer to two point mutations of a protein, two splice variants or even two somewhat distantly related genes. Actually, **a word that we could get rid of, for more clarity in the scientific literature**.

Isoprenoid: A class of natural substances that in the LEGO kit of life is assembled from only two C5 building blocks. Later modifications allow a virtually infinite variety of compounds. Plants are probably the isoprenoid synthesis world champions.

Library: Something where a collection of complex biological things is inside. This can be a substance library, in which possible drug precursors are searched for. In the context of display techniques, this refers to a collection of DNA or RNA molecules that code for possible binding molecules (aptamers, antibodies). It is helpful to be able to estimate the actual size of the library.

Ligand: An alloy (of metals) comes from the Latin word *ligare*—to bind. Just like the alloying of a fine gravy sauce. And there we also have the cute word ligand. A ligand is pretty much anything that binds to a protein (or another biopolymer), a binder, usually smaller than that, to which the ligand binds. That is then the target molecule or the receptor.

Liquid-liquid phase separation → Phase separation

Local energetic minimum → Prebiotic chemistry

Monomer → Polymer

Mutation: A change in the sequence of coding nucleobases (usually DNA, sometimes RNA). Depending on the context of the mutation and what it ultimately causes, different types of mutations are distinguished. Within sequences that code for proteins, a different codon can arise, but it stands for the same amino acid—a *silent* mutation. Then there can be changes in the amino acid sequence—a *missense* mutation. Finally, a stop codon may arise, causing a *non-sense* mutation.

Next-Generation-Sequencing → Sequencing (of DNA)

Orbital: Orbitals of atoms or molecules are wavering electron clouds or also probabilities of electrons' presence. A bit like a vibrating string of any string instrument, orbitals can take on strange shapes—sometimes they are round, sometimes they resemble a dumbbell from the fitness studio, sometimes they have even more complicated shapes. Biochemically, s-, p- and d-orbitals are important as well as various mixtures of s- and p-orbitals; the sp^3—and sp^2 -hybrid orbitals. → Stereochemistry

PDB → Protein Data Base

PDB-Format → Protein Data Base

Phage-Display → Display

Phase separation, Phase transition: When it comes to water, this primarily refers to the melting of ice and the freezing of water as well as the evaporation of water and condensation of water vapor. Biochemically, however, these can also be other phase transitions, such as in a membrane from a crystalline to a liquid state. And then there is the wide field of liquid-liquid phase separations. These are being discovered more and more in all sorts of biochemical processes. Think of the absolutely biopolymer-packed interior of a cell—all in aqueous solution—then however lipids or nucleic acids, or something else, aggregates, phase separation occurs, and suddenly that part of the cell behaves like oil, from now on.

Polymer: "-mer" is a very important syllable in biochemistry, it comes from the ancient Greek *meros*, which means part or share. So, let's count: Monomer = something made of one part, usually something small, chemically often also something reactive; Dimer = a two-part substance; Tetramer = four-part; and finally, Polymer = a thing made of many (same?) components.

Primordial Soup Experiment → Prebiotic Chemistry

Protein Data Base: The PDB is an international repository for 3D structures of biopolymers, established in 1971. It has mirror sites in different parts of the World. Most of the structures in it are structures of proteins. The PDB also contains structural information from numerous ligands, lipids and cofactors, as well as DNA and RNA complexes. PDB stands simultaneously for the PDB format, in which the structural information is stored. PDB files can be visualized with various PDB viewers or browsers.

Protein: We have talked a lot about proteins, built from amino acids, and already have mentioned that proteins as a compound class are named after the Latin word "proteus". In some languages, proteins alternatively are named after egg white, resulting in something like "albumin"—the *white stuff*.

Receptor: In endocrinology, a receptor is exactly what a specific hormone binds to, thereby unfolding its biological effect. In pharmacology, this concept is much more vague—there it is everything that a drug binds to in order to unfold its pharmacological effect.

Reporter → Indicator

RNA World // Ribozyme // Ribosome: The three R's of prebiotic chemistry. The RNA world hypothesis attempts to explain the beginning of life when there was neither DNA as an information store nor proteins as biocatalysts.

Schrödinger's Cat: Few thought experiments are better known than the cat of physicist Erwin Schrödinger. In this thought experiment, Schrödinger illustrates the difficulties of interpreting too directly what you might encounter in the quantum world. It's about the fact that two states, which are otherwise not compatible with each other, can indeed be superimposed in the quantum world. Perhaps comparable to that time after handing in a difficult exam, but before the grades are announced—a strange time in which you somehow passed the exam, but have failed them, at the same time.

Second Messenger: The transmission of a message from outside into the cell is a science in itself. Once the signal is in there, the signal transmission with second messengers begins. These signal molecules are produced within the cell in response to a stimulus. Much-discussed second messengers are the cyclic nucleotides cAMP and cGMP or calcium ions. There are many others. How specificity is ensured in the whole process, is the question that almost always arises in the end.

Secondary Metabolism: If you say A, you must also say B. The central metabolism is something like making proteins and sugars. Everything that goes beyond that, for example glucosinolates and isoprenoids, is then the secondary metabolism. This is how the term secondary metabolism could have originated. In green biotechnology and plant physiology, an established term, it perhaps is even applicable to animals or humans.

Self-Assembly: Order from chaos. Smaller subunits of larger complexes assemble themselves; usually driven by many small interactions and/or packing effects.Very chic self-assembly systems are viral capsid proteins—even in thin, aqueous buffers they can (sometimes) form complete viral shells. → Icosahedral sphere packing, densest

Sensor → Indicator

Sequencing, **Next-Gen-Seq**: The sequencing of nucleic acids (almost always DNA) has now reached an industrial age. Sequencing-by-Synthesis is a principle that has been massively parallelized. Nanopore sequencing is sequencing in a completely different way. Here, a base-specific electrical voltage is generated while a single strand of DNA is pulled through a special protein pore. Due to all sorts of methodological advances, sequencing has become incredibly fast. At the beginning of 2022, the world record for the complete sequencing of a human genome was 5 hours and 2 minutes.

Sequencing: First of all, sequencing is the determination of the sequence of any linearly composed biopolymers. So sugar chains or peptides could also be sequenced. More specifically, it refers to (i.e., in 99.999% of cases) the sequencing of DNA. The standard procedure for this is di-deoxy sequencing according to Frederick Sanger. → Next-Gen-Seq

Signal Transmission: A living organism must constantly exchange information with the outside world—and this was the case even before the times of smartphones. There is such a flow of information between different organs and/or cells. Often, a limited number of chemical signal molecules are used for this purpose, which we call hormones. Once arrived at the recipient, such a hormone docks onto a corresponding receptor, which then stimulates further signal transmission within the cell through various second messengers.

Sphere packing, densest: A very good starting point, to understand self-assembly. Spheres loosely poured out and then shaken a bit will always assume the densest sphere packing. The sphere shape dictates this arrangement. Somehow comparable to the round diameters of phospholipid heads, when they form a biological double membrane.

Stereochemistry: Regardless of whether the molecule is small or large, we usually see it on a flat screen or printed on paper. Stereochemistry is two things at first. I can use the knowledge about the real geometry of a chemical compound to see this molecule in three dimensions—as if I put on some sort of learned 3D goggles. At the same time, I can also use tools to achieve the same goal. These tools can be Magic Eye images or molecule model building kits. Programs that display three-dimensional models of protein structures are also great. Oh, and then there are 3D printers and such neat software that can represent a mix of simulation and reality; we have arrived at augmented reality.

Sulfation: It starts so harmlessly: Just make an ester between sulfuric acid and any biomolecule with an OH group (alternatively an acid amide with a molecule with an amino group). Biochemically, the trouble then starts with the reaction-inert sulfate—an enzyme that helps the sulfate with the aid of ATP provides a remedy here. Sulfating enzymes appear in many specialized metabolic pathways. Sulfation changes the functioning of steroid hormones and many other biomolecules. And then there is the selective desulfation in peripheral tissues. Ah, and all this was only the cytoplasmic sulfation events. In the Golgi, tailor-made sulfated glycosaminoglycans are then made, which in turn bind various growth factors.Sulfation remains an exciting subject.

Surface tension: In biochemical comparison, water has a very high surface tension. This is because water likes to stay among itself, often quite disordered. At the surface, however, water is forced to adopt a certain order. Therefore, water always wants to adopt a shape with the smallest possible surface per volume. By the way: raindrops are round; they never have the "typical" raindrop shape—never.

Target molecule: The molecular target for any intervention; usually used in the context of drug development. The target molecule is the structure in the body to which the active ingredient in a drug will dock to, to unfold its pharmacological effect. An example: The target molecules of the active ingredient Sildenafil in the oddly shaped blue Viagra tablets are the enzymes phospho-di-esterase 5 and 6.

Tetrahedron: An important geometric shape to understand the bonding geometry of the elements carbon, nitrogen, and oxygen. They are all elements of the second period of the periodic table. A methane molecule consists of one carbon atom and four bonded hydrogen atoms. All bond angles are exactly the same; the perfect bond angle in the tetrahedron is 109.5 degrees.

Tetramer → Polymer

Thermodynamic equilibrium → Thermodynamics

Thermodynamics: A very large theoretical structure of chemistry. Biochemically, it often means what one could get out of a chemical reaction, maximally or theoretically. Alternatively, whether a reaction would proceed voluntarily, or whether one would have to support it with ATP hydrolysis, or similar. Thermodynamic measurements usually assume an equilibrium state, that (hypothetical) moment when everything has been said, and all forward and backward reactions no longer cause any change in the composition of substances.

Titration: A word that biochemists dearly love. A titration or the process of titrating something is about determining its concentration and/or its composition. In a pH titration, one determines the acid strength, also known as the pK_A value, of a compound or a functional group—the mid-point of the process of giving away or accepting a proton. A quantitative research method, lots of fun.

Trigonal Bipyramid: An important geometric shape to understand the bonding geometry of the elements carbon, nitrogen, and oxygen. They are all elements of the second period of the periodic table. Trigonal bipyramids are nice to look at. A central atom and three bonded atoms all lie in one plane; above and below this plane, a free pair of electrons wafts or orbits around in a dumbbell-shaped p-orbital.

Uncoupler protein: At first it sounds nasty—a protein that ruins all the beautiful work of the OxPhos respiratory chain and makes the proton gradient shrink like a (carefully) pricked balloon. But then it becomes clear that the uncoupler protein and a few other mechanisms generate heat without muscles having to work for it. If one just had more brown fat tissue, fat would simply melt away.

Volume-Surface Inequality: Small things are not just simply smaller than large ones; sometimes they function in quite different ways. When a living thing increases in size, its mass or volume grows with the third power (cubic), while the surface only grows linearly. Therefore, a tiny creature has very different "worries" than a very large creature. First of all, this is a "mere" mathematical dependency. However, one or another biological rule can be derived from it, for example the rule of Bergmann, according to which the body size of closely related species increases towards the polar regions of our Earth.

Z-DNA: One of several conformations of DNA. It is thought to be created when overtly underwinding DNA, to release torque. Frankly, I have listed Z-DNA in this glossary only for having a word that starts with Z, to end with.

The manufacturer's authorised representative in the EU is Springer Nature Customer Service Centre GmbH, Europaplatz 3, 69115 Heidelberg, Germany. If you have any concerns regarding our products, please contact ProductSafety@springernature.com

Printed and bound by CPI Group (UK) Ltd, Croydon, CR0 4YY
07/07/2026
02160920-0001